Running Silver

Also by John Waldman

Stripers: An Angler's Anthology

Heartbeats in the Muck: The History, Sea Life, and Environment of New York Harbor

The Dance of the Flying Gurnards: America's Coastal Curiosities and Beachside Wonders

100 Weird Ways to Catch Fish

Still the Same Hawk: Reflections on Nature in New York

Running Silver

Restoring Atlantic Rivers and Their Great Fish Migrations

John Waldman

LYONS PRESS
Guilford, Connecticut
An imprint of Globe Pequot Press

Lyons Press is an imprint of Globe Pequot Press.

Support provided by Furthermore, a program of the J. M. Kaplan Fund.
Song lyrics on page 6 from "Waters of March" © 1972 by Antonio Carlos Jobim.
Fish silhouettes licensed by Shutterstock.com.

Map on page viii by Melissa Baker © Morris Book Publishing, LLC
Project editor: Meredith Dias
Text design: Sue Murray
Layout: Joanna Beyer

Library of Congress Cataloging-in-Publication Data

Waldman, John R.
  Running silver : restoring Atlantic rivers and their great fish
migrations / John Waldman.
     pages cm
  Includes bibliographical references.
  ISBN 978-0-7627-8059-4 (hardback)
  1. Migratory fishes—Atlantic Coast Region (U.S.) 2. Migratory fishes—Atlantic Coast
Region (U.S.)—History. 3. Fishes—Migration—Atlantic Coast Region (U.S.) 4. Fishes—
Ecology—Atlantic Coast Region (U.S.) 5. Fishes—Conservation—Atlantic Coast Region
(U.S.) 6. Environmental degradation—Atlantic Coast Region (U.S.) 7. Rivers—Atlantic
Coast Region (U.S.) 8. Stream restoration—Atlantic Coast Region (U.S.) 9. Atlantic Coast
Region (U.S.)—Environmental conditions. I. Title.
  QL621.5.W35 2013
  333.95'6097—dc23
                                                              2013024382
Printed in the United States of America

10 9 8 7 6 5 4 3 2 1

For Thoreau, who could hear the fishes cry

Contents

Map viii
Preface: Who Hears the Fishes When They Cry? ix
Prologue: Dammed-Nation xi

  1   Running Silver and Ghost Fishes 1
  2   Diadromy 101: Swimming the Great Migratory Circuit 6
  3   The Seasonal Parade 18
  4   On the Nature of Rivers 48

Interlude I   A Shad's Journey, circa 1600 59

  5   On Natural Abundances: Remembering Not to Forget 63
  6   Spearfish Moon 72
  7   Providence and Plenitude 83
  8   Floating Caskets and the Pennsylvania Navy 96
  9   Billions of Fish in Hot Water 106
10   Precautionary Principle vs. Principally Not Cautious 113
11   Concrete Crimes against Rivers 124
12   Climate Change: Latitudes and Attitudes 141
13   Migration and the Exotic Species Gauntlet 156
14   Giants of the Rivers: Gone Forever? 167
15   Peering into the Black Box 176
16   Hatchery Stocking: Subtraction by Addition 186
17   Dam Removal: Fish vs. Ignorance and Inertia 202
18   Fish Passage, or Not 212

Interlude II   A Shad's Journey, circa 2013 229

19   Favorable Currents, Fortunate Confluences 234
20   Toward a New Stewardship 253

Epilogue: Keep a Stiff Fin 260
Acknowledgments 261
Endnotes 263

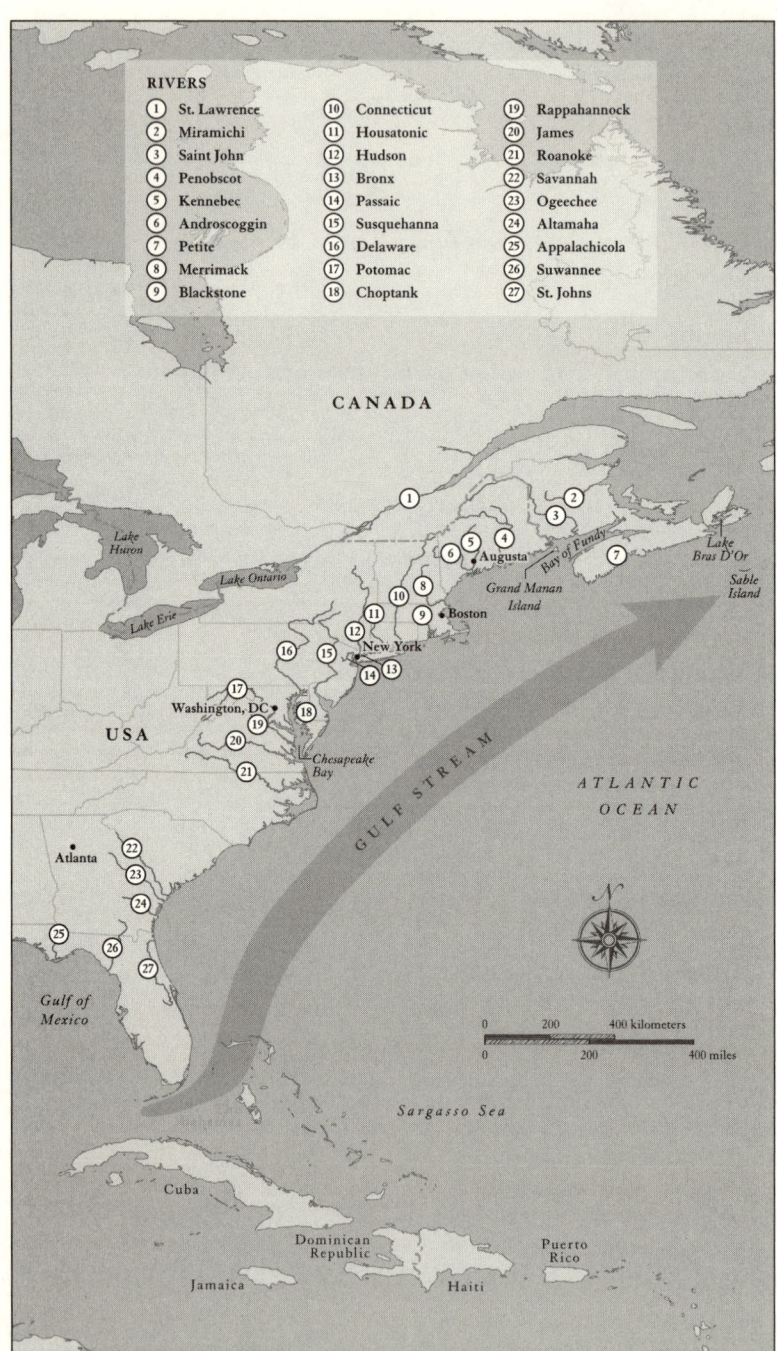

RIVERS

①	St. Lawrence	⑩	Connecticut	⑲	Rappahannock
②	Miramichi	⑪	Housatonic	⑳	James
③	Saint John	⑫	Hudson	㉑	Roanoke
④	Penobscot	⑬	Bronx	㉒	Savannah
⑤	Kennebec	⑭	Passaic	㉓	Ogeechee
⑥	Androscoggin	⑮	Susquehanna	㉔	Altamaha
⑦	Petite	⑯	Delaware	㉕	Appalachicola
⑧	Merrimack	⑰	Potomac	㉖	Suwannee
⑨	Blackstone	⑱	Choptank	㉗	St. Johns

Map of rivers referred to in the book

Preface
Who Hears the Fishes When They Cry?

I was an odd child—an urban Bronx boy who grew up near the salt waters of Long Island Sound but who at a young age became haunted by rivers. Early memories include a family picnic forty miles north of New York City where I was smitten with the life in a small trout stream, marveling at schools of minnows and dace in spawning colors that courted and darted over the bottom of a winding meadow reach. And of going to the curb in front of my house one winter day when rushing snowmelt created pools of water in the icy gutter, and my moving rocks around to make ponds and rapids, imagining migratory fish pushing up a mighty river. Later in life I had the opportunity to get to know some of the world's major flowages—Mongolia's Ur, Ukraine's Dnieper, and Alaska's Copper—but I also experienced the pleasures of the diminutive Bronx River in my own metaphorical backyard.

I love water in all its forms, and can't get enough of the open ocean, the seashore, mangrove jungles and salt marshes, and weedy ponds and ice-covered lakes, but I am still haunted by rivers. I've spent my career delving into the mysteries of these migratory freshwater-sea fishes, the work interrupted whenever possible by also probing for them with hook and line. I've been privileged to enjoy these highly direct forms of learning, something increasingly rare in the modern world, which has added to my persevering sense of awe.

I remain taken with the marvelous resilience of these fishes—if only offered a reasonable chance—but I am deeply worried about their prospects. And I believe their potential to help feed us, to contribute to the healthy functioning of our ecosystems, and to help renew our spirits with knowledge of their grand life histories and by viewing their miraculous returns to rivers in springtime is being lost to a tangle of antagonistic forces, among them, the erosive power of time on memory, ever since they existed in astonishing plenitude. This book follows these fishes through the river of history to understand how they have arrived at this now sorry state, and how we might reverse their fortunes and, in the process, help ourselves.

Thoreau, so dismayed by what he foresaw for these fishes during his journey on a rapidly industrializing river, asked, "Who hears the fishes when they cry?" Maybe by revealing a squandered legacy, more people will begin to listen.

Prologue
Dammed-Nation

Riddle: What did the fish say when it swam into the wall?
Answer: Oh dam!

I'm on a strange mission as I drive through the South Bronx. Fish in this part of the world usually means *branzini* in the nearby Little Italy neighborhood, or *baccala* in the many local Hispanic bodegas. But I've got two of my fish-obsessed graduate students in the car—George Jackman, a forty-five-year-old former New York City policeman who knows this section of town from the front of a squad car and the back of a police horse, and Peter Malaty, a twenty-one-year-old wild-haired angling fanatic who spends nights on rocks in storm-tossed waves, casting for striped bass. With those bona fides, neither is worried about the South Bronx's reputation for crime.

We pass through the West Farms neighborhood, where arson ravaged its buildings in the 1970s, and where even today students pass through metal detectors to enter school, and we pull into River Park, just downriver of the first dam on the Bronx River. Cup your hands to your eyes to block your peripheral vision and this view of the river looks downright pastoral. The Bronx River isn't much of a watercourse by world standards, but it is spectacular for New York City, which is enveloped by the great Hudson Estuary but which lacks classic fresh running waters like this. Nonetheless, don't take the Bronx River too lightly; in July 2010 a teenage girl swimming illegally in the deep plunge pool below the dam drowned, as did the teenage boy who tried to save her.

We're at the river this April morning in 2009 to set a fyke net in the flow to catch alewives on behalf of the New York City Parks Department. Alewives are foot-long silvery fish that resemble the much larger American shad, both of which swim into rivers from the sea in spring to spawn. Once, virtually every temperate Eastern Seaboard river and stream had runs of alewives, sometimes numbering in the millions. Lately, though, it would have been hard to take the biological potential of the Bronx River seriously. Flowing with a milky hue down the middle of a borough with almost a million and half people, it passes quietly behind businesses and urban parks while seeming to have a remarkable number of discharge pipes aimed at it before it enters a long ponded section behind the Bronx Zoo. Still, a striking signal of recovery was seen in 2007 when a beaver swam down the river from the northern suburbs and made a home on the wooded banks of the zoo property. Named José, after Bronx congressman José Serrano, the lone animal was joined by another in 2010, promptly doubling the Big Apple's wild beaver population.

The reason we were looking for alewives was that three years earlier, to great media fanfare, Connecticut state biologists, led by fish restoration guru Steve Gephard, drove a specialized tank truck to deliver to the Bronx River ready-to-spawn alewives netted from a brook in eastern Connecticut. I was there that day in May 2006 and watched Congressman Serrano preside with apparent delight as some two hundred frisky adult alewives washed down a

Congressman José Serrano stocking alewives into the Bronx River, May 2006
Julie Larsen Maher © WCS

tube into the pond at the zoo. One year later another four hundred potential spawners were added. No one verified directly that successful reproduction occurred, but there was a reliable sign. Young alewives before they migrate to sea in late summer make a characteristic "popping" splash at the water surface, and popping was seen upriver of the dam for the first time ever. The alewives appeared to have spawned—but would they return? Though 2011 was when most would be expected to come back to the river if the experiment worked, 2009 was the first year where a few conceivably might be seen.

Fyke nets are complicated affairs, a series of chambers held erect with hoops headed with wing nets opened wide to steer upriver migrants into the hold. We staked the contraption to the river bottom below the dam and left it to do its job.

The Amoskeag Fishways Learning and Visitors Center in the old industrial city of Manchester, New Hampshire, is something of a tourist attraction. Drive the roads along the Merrimack River on either bank and you will spot a giant sculpture of an American shad—a glorious, blue, leaping facsimile of this wild creature from the sea—that marks the Center's location behind the Amoskeag Dam. The Center advertises itself as an award-winning environmental education center and includes an impressive exhibit hall that is open year-round. It even invites you to "Celebrate the Magic of the Merrimack."

The Merrimack River once was home to enormous runs from the sea of alewives, Atlantic salmon, Atlantic sturgeon, and sea lamprey, in addition to American shad. *Amoskeag,* a Native American word, means "great fishing place," but the high potential energy of its rushing waters was harnessed in the 1800s for power for the Merrimack Valley's many textile mills. The Amoskeag Dam, now the third dam from the sea, was built in 1836. Twenty-two-year-old Henry David Thoreau paddled the Merrimack and its tributary, the Concord, in the summer of 1839 with his brother John and published his observations on a watershed experiencing rapid transformation in his *A Week on the Concord and Merrimack Rivers.* Lightly regarded and little read then, the book is considered prescient now. Thoreau frequently commented on his fears for the fate of the Merrimack's migratory fishes. Describing the effects of a downstream dam, he wrote, "Salmon, Shad, and Alewives were formerly abundant here, and taken in weirs by the Indians, who taught this method to the whites, by whom they were used as food and as manure, until the dam, and afterward the canal at Billerica, and the factories at Lowell, put an end to their migrations hitherward."

Hitherward soon became afterward. A New Hampshire government directive forced the Amoskeag Dam owners to build a fish ladder to help these migratory fish move farther upriver, but it was a failure and collapsed by 1850. In 1921 the dam was rebuilt to provide hydropower, but it wasn't until 1989 that a modern, fifty-four-step fish ladder was installed to fulfill the dream of allowing the migratory fish to reach their primordial spawning grounds. The visitor center followed one year later. One of its chief attractions was an underwater glass viewing window to allow the public to see all the shad and other migratory species that would march upriver.

What have visitors actually seen through this looking glass? Restoration targets for anadromous fish runs in the Merrimack are modest for such a large river: 300 Atlantic salmon to simply reach the first dam on the system, and 35,000 shad and 300,000 alewives (and its close relative, blueback herring) to actually pass it. On average, though, only 16 percent of the fish that pass the first dam also pass the second. Between 2005 and 2010, the number of alewives and bluebacks enumerated at the first dam averaged fewer than 800 individuals. With such dismal numbers, any actually reaching the Amoskeag Dam should be a cause for celebration. Here are the counts from 2012: The public viewed exactly zero Atlantic salmon, zero alewife or blueback herring, zero striped bass, zero sea lamprey, and one American shad—but only if they looked up and saw the sculpture as they drove in.

Thoreau felt deeply for the shad's plight, and no doubt would be dispirited by the Merrimack today. He wrote that shad were "armed only with innocence and just cause," and that he was at one with them; he even foreshadowed dam removal with "and who knows what may avail a crow-bar against that Billerica dam?"

On July 1, 1999, a crowbar was put against another major dam, and its collapse still echoes through the river conservation world. Built in 1837, the Edwards Dam in Augusta, Maine, was a 25-foot-high, 900-foot-long, timber-and-concrete wall that stopped all of the Kennebec River's anadromous fish from reaching their ancestral spawning grounds. As for other major northeastern US rivers, these fish runs were legendary. In 1723 a French priest described alewives moving up the Kennebec "in such numbers that a person could fill fifty thousand barrels in a day, if he could endure the labor." Like the Merrimack, the Kennebec did not escape the Industrial Revolution, leaving a legacy of barriers throughout its watershed. But the next dam upriver was seventeen miles distant, which made the Edwards a tempting target for removal. If accomplished, not all—but at least a substantial reach—of the original spawning grounds could be reclaimed. With it being a hydroelectric dam, though, it didn't seem possible. It was regulated by the Federal Energy Regulatory Commission (FERC), a government body notoriously friendly to the industry.

Still, the dam's license was coming up for renewal in 1997, and the climate for taking down dams was starting to heat up. Environmental activists sensed an opening and went for it, pressing a 1986 federal law that requires FERC to "balance the environmental impact of a dam against the value of the electricity it produces." The Edwards Dam was not a major energy producer; the 3.5 megawatts it produced only filled one-thousandth of the state's electricity needs. Or, as John McPhee wrote in *The Founding Fish,* barely enough to light L.L. Bean's warehouse. But it was a major fish stopper.

The case was made. Never before had FERC denied a renewal license for a hydroelectric facility on a major river. Information on all aspects of the dam's costs and benefits were submitted, including accounts of historical fish runs. Even the fish cooperated with karmic public displays. Ronald Kreisman, a hard-driving environmental lawyer with a soft spot for rivers, remembers taking regulators to the site during the spring, when the river's relict migratory fish stocks were aggregated below the dam. Once, with visitors on board a boat, Atlantic sturgeon continuously leapt from the water, the airborne fish seeming to plead for remediation. Another time, walking near the base of the dam, the water boiled with the silver bodies of frustrated shad and alewives.

Somehow reason won. The dam came down. John McPhee was there on July 1, 1999, and described a crowd in a giant parking lot just downriver from the dam: people in suits, ties, and combat fatigues, sandals, sneakers, boots, and backpacks. There were T-shirts for sale, booths of brochures from agencies, television cameras, a helicopter, small airplanes, live music, and speeches by many, including the mayor of Augusta, the governor of Maine, and Bruce Babbitt, Secretary of the Interior. This was a *happening.* Grassroots, bottom-up inspired, like so many removals of smaller dams, but this time also top-down blessed. Perhaps more importantly than the gathering of human migrants, that morning schools of striped bass had been seen nosing up to the dam, and a salmon broached two feet out of the water, seeming to just hang in the air as if watching.

As Maine goes, so goes the nation?

George and Peter checked the fyke net daily. A few suckers and other river resident freshwater fish showed, but no alewives. A week later, however, George burst into my office as if he'd collared one of America's Ten Most Wanted, shouting, "We caught alewives!" That morning they'd seen silver in the net—two healthy-looking alewives fresh from a short trip through an industrially ravaged tidal creek that connects this section of the Bronx River with salt water. This was exciting news to the New York City environmental community. Perhaps a fish ladder could now be built to allow alewives to reach the long pond behind the zoo—what appeared to be prime spawning habitat. But would it work?

Ideally, from the point of view of a healthy river, the dam should come down. The Edwards Dam removal had worked; almost instantly, salmon, shad, striped bass, alewives, and other anadromous fishes pushed up the reopened seventeen miles of river and their populations began to increase. But removing this dam on the Bronx River was not on the table. Would a fish ladder here be another Amoskeag Fishway—an expensive device with every bell and whistle except fish? I couldn't help but think of how the Amoskeag Fishways Learning and Visitors Center encourages visitors to "learn the importance of wise stewardship of our water resources." It does so with no apparent intention or seeming awareness of irony.

Not that it's my choice to make, but because the stocking of alewives seemed to have worked, I decided the fish ladder was worth trying. Although fishways on many large rivers perform poorly, alewives often respond well to ladders on small dams sited not far from the sea. Just recently, twenty miles east of the Bronx, a ladder that went up on the Mianus River in Connecticut raised the alewife and blueback herring run size from about 5,000 the first year to around 100,000 a few years later.

My ever-curious colleague Karin Limburg practices due diligence, though, and asked to look at the ear bones or "otoliths" of the returning alewives. Otoliths form annual bands like the rings of a tree and are more reliable for aging a fish than the traditional method of counting bands on scales. Karin sectioned the bones, inspected them under a microscope, and calmly announced that the alewives were five years old—older than is possible for the progeny of any of the stocked fish. This unlikely, quintessentially urban setting had its own relict alewife run, but no one had bothered to look. They'd been there all along!

Chapter 1

Running Silver and Ghost Fishes

The herring are running!

—*John Hay,* The Run *(1959)*

Henry David Thoreau was not pleased with the changes he saw occurring along the Concord and Merrimack Rivers as he and John paddled. The voyage that was the basis for *A Week on the Concord and Merrimack Rivers* actually took two weeks, not one. Each chapter in the book constituted a "day," but was really a compendium of observations made along the rivers, together with ruminations on religion, poetry, and history. Thoreau may be celebrated for his philosophical insights, but he was also an acutely perceptive observer of the natural world. In fact, his data on the timing of the flowering of Massachusetts flora are being used as a baseline for measuring climate-change effects today. In *A Week,* Thoreau foretold many of the oncoming environmental consequences of the nascent Industrial Revolution on the rivers he so admired. The book's first draft was written while Thoreau lived at Walden Pond. When completed, he could not find a publisher for it, and so he had it produced at his own expense. *A Week* was not a smash success; only a few copies were sold of the hundreds Thoreau had printed, and he was driven into debt.

Thoreau was familiar with the great abundances of anadromous fishes as they drove inland, and it's possible he witnessed a migration so pronounced as to justify use of the term *running silver,* a description of times when there were so many metallic-scaled bodies churning their way up a river that it seemed the fish had become the water. I have collected testimony, written accounts of the reactions of early colonists to these runs when they ran silver. The language often approaches the hyperbolic, as if the possibility of such abundances could not have been conceived, never mind witnessed. To pore over these quotations is to read of awesome plethora, number orders of magnitudes higher than we are accustomed to today.

A sampling:

We are set down eighty miles within a river, for breadth, sweetness of water, length navigable up into the country, deep and bold channel, so stored with sturgeon and other sweet fish as no man's fortune has ever possessed the like.

Yea, when a heape of stones is reared up against [the alewives during their spawning runs] a foot high above the water, they leap and tumble over and will not be beaten back with cudgels.

We had more sturgeon than could be devoured by dog or man . . .

In the spring of the year, herrings come up in such abundance into their brooks and fords to spawn that it is almost impossible to ride through without treading on them. Thus do these poor creatures expose their own lives to some hazard out of their care to find a more convenient reception for their young, which are not yet alive. Thence it is that at this time of the year, the freshes of the rivers, like that of the Broadruck, stink of fish.

[T]he greate smelts passé up [the Smelt River, near Plymouth, Massachusetts] to spawne likewise in troupes innumerable, which with a scoupe, or a boule, or a peece of bark, a man may cast upon the bank . . .

There are such multitudes, that I have seene stopped into the river close adjoining to my house with a sand at one tide, so many as will loade a ship of a 100 Tonnes. Other places have greater quantities in so much, as wagers have bin layed, that one should not throw a stone in the water, but that hee should hit a fish. I my selfe at the turning of the tyde, have seene such multitudes passé out of a pound, that it seemed to mee, that one might goe over their backs drishod.

The sturgeons be all over the country, but the best catching of them is upon the shoals of Cape Cod and in the river of Merrimac, where much is taken, pickled, and brought for England. Some of these be 12, 14, 18 foot long. I set not down the price of fish there because it is so cheap . . .

[D]uring one month the fish ascend the river in so great numbers that a man could fill fifty thousand barrels with them in a day, if he could be equal to the work.

And my favorite, a veritable festival of superlatives:

When they spawn, all streams and waters are completely filled with them, and one might believe, when he sees such terrible amounts of them, that there was as great a supply of herring as there is water. In a word, it is unbelievable, indeed, indescribable, as also incomprehensible, what quantity is found there. One must behold oneself.

This sampling extends from 1607 to the early 1800s. They reflect the perceptions of European colonists who had recently arrived in the New World, from an Old World where its fishes had been overharvested for centuries. Their view of "normal" abundances was already altered by the insidious declines long suffered by their fish stocks. Yes, Native Americans had fished the American runs for millennia, but to help feed human populations that were diminutive compared with the numbers of people in Europe. Walking "drishod" over the backs of migrating fish is, of course, an exaggeration, but one expressed in wonderment by colonists who were overwhelmed, both with the sheer spectacle and with their own good fortune.

One December day in 2007, my colleague Karin Limburg and I sat with our laptop computers in her dining room in Syracuse, New York, an ad hoc war room, to try to take stock of the status and trends of the Atlantic's diadromous fishes. Both of us had worked on these species long enough to know we'd end up painting a grim picture, but until then, no one had synthesized so much of this bad news quantitatively. An early snow fell hard outside as we began to merge diverse streams of information.

The news was not good; in fact, it was downright dismal. The resultant article, published in the journal *Bioscience,* covered two dozen species of fish. We had examined their abundances as far back in time as possible, realizing that the numbers they would show likely were already reduced, probably substantially in many cases, from the pristine profusion seen before anyone was inspired to try and enumerate them. Nonetheless, the declines were astonishing: For thirty-five species, each considered as a whole, or of populations of a species, relative abundances had dropped more than 98 percent, from historic peaks in thirteen, and more than 90 percent in another eleven. Most had reached their lowest levels at the present, and many showed trend lines that sloped slowly toward zero.

Many individual populations had been lost, too. In North America for Atlantic salmon, only 135 of some 600 original runs were left. American shad were extirpated in almost half the rivers where they once occurred. The Atlantic whitefish lost only one population, but it only had two to begin with, and it teeters near extinction. In Europe the sea sturgeon, which had swum in as many as twenty rivers from the Baltic all the way to the Black Sea, hung on only in France's Gironde River, and in grim numbers. For some other species significant losses of populations are suspected (such as sea lamprey), but because they lack commercial value, no one had bothered to collect the appropriate data. The conservation status of these fishes echoed our more specific data. The International Union for the Conservation of Nature convenes panels that deliberate at length about the state of a species; most were now listed with a designation of at least some concern, or worse: "vulnerable," "near threatened," "endangered," and "critically endangered."

The only encouraging news Karin and I extracted is that true species extinctions haven't yet occurred. Overall take-home message: *Abundances for most species decimated, numbers of populations sometimes severely reduced, but flirtations with extinctions still uncommon.* A set of findings that simultaneously seem even beyond the point of urgency while revealing a modicum of hope.

Whether as severely diminished numbers or as populations lost, what's happened to these freshwater-sea migratory fishes has left both ecological and sociological voids: *ghost fishes*. "Ghost species" is a new concept in conservation biology. Concerning the freshwater-sea migratory fishes, ghost fishes are those that are either completely lost as a population in a river or region where they once occurred, or those that persist at such low numbers that in ecological terms, they are essentially absent. Though now missing in reality or in effect, these fishes once did play a role, and usually a highly important one, in the broader food web as prey, predator, and competitor, one that co-evolved with the other organisms that compose that network. And so their absence resonates as holes, or "ghosts," in the ecological machinery of those environments.

In the absence of these fishes, their importance to society naturally faded, also to ghost-like roles. The task today is to exorcise these ghosts, not through the supernatural but by filling the empty spaces in nature they represent through the hard work of applied restoration via all possible avenues. But to muster the wherewithal, their fates first need to matter; in our minds, they need to pass from poorly remembered specters to living creatures in need of a fair chance. Or, as Thoreau put it, "Poor shad! Where is thy redress?"

Chapter 2

Diadromy 101: Swimming the Great Migratory Circuit

[A]nd the riverbank talks of the waters of March,
it's the promise of life, it's the joy in your heart.
—*"Waters of March," Antonio Carlos Jobim
and Cassandra Wilson*

Alaska at last! I rejoice as an East Coast conservation biologist and angler finally in salmon paradise. My buddies and I have arrived in the quaint coastal village of Cordova too late in the day to go fishing, but we can't help but drive at last light to a small stream to look at a spawning run of pink salmon. Everywhere in the current the dark backs of the pinks are showing above the surface as fish either pass farther upstream, dig pits in the gravel called "redds" with their tails to mate in place, or just loiter. Others appear less lively and actually "torn," white flesh showing as these spawned-out individuals literally rot alive as preprogrammed death advances. And dead salmon line the banks, decaying and leaving a powerful but not completely unpleasant organic stench in the air.

I want to touch and handle a salmon, so we find a curved tree branch and steer a female pink onto shore. I hold my pink lady high to admire her and some large golden eggs trickle out, so I put my mouth to her vent and taste a few—they burst and provide me with the subtle and salty flavors of a life spent wandering the Pacific. When I slip her back into the flow, she immediately rejoins the salmon parade. On another day, while driving I look down along the shore and spot a run of pink salmon streaming through a culvert. I stop the car and see that the landward side of the road has a small flow in its gutter and it is filled with salmon with their backs out of the water. Where they are headed amazes me; a short distance away the source of this runoff is nothing but a trickle down a 45-degree rock-strewn slope. But the pinks are ascending from wet pocket to wet pocket like drunken mountain climbers, taking a harsh physical beating to place their eggs in the relative safety of this otherwise inhospitable environment.

I've spent much of my career in New York working on regional and international issues concerning freshwater-sea migratory or "anadromous" fishes, but no sight ever revealed so baldly the sheer force of the spawning drive as in these Alaska fishes. This is anadromy illustrated, and I am in awe. And yet, I realize it's no different for their Atlantic Coast analogues; the combination of this instinctual impulse plus overwhelming plenitude is just either more cryptic, or it's a legacy that's been squandered through mismanagement and obliviousness. A 300-pound sturgeon that slides past Manhattan to spawn in the Hudson is responding to the same imperative; it's just that the fish is forty feet below in a murky river. Likewise

A Hudson River commercial fishing station. Note that all fish species shown are anadromous, demonstrating their high importance.
From The Hudson: From the Wilderness to the Sea *(Benson Lossing, 1866)*

for an American shad in Georgia's Ogeechee, or a striped bass in Maine's Kennebec. In fact, most rivers of the Atlantic Seaboard did run silver with alewives and other fishes before the Industrial Revolution, obeying the same imperative; it's just that those days are long gone and barely remembered.

There are more than 20,000 species of fish alive on the planet today. Imagine an experiment: Take one of each of them and slowly change their dwellings from freshwater to salt water, or salt water to freshwater, to see how many survive. The overwhelming majority should perish—most when crossing the barrier of about five to ten parts per thousand salinity from either direction. Of those that can make the physiologically demanding transition between the zero parts per thousand salinity of freshwater and the thirty-five parts per thousand of seawater (or vice versa), a select group of only about 250 do this as a routine and predictable part of their life cycle. These are the *diadromous* fishes.

Some more essential terminology: *Anadromous* fishes are those diadromous fishes that are spawned in freshwater and then migrate to the sea. *Catadromous* fishes do the opposite; they start life in the ocean and migrate to freshwater. A little more "Intro to Ichthyology": "Fish" are the individuals you have in your bucket after your successful fishing trip. Perhaps you had a good day on the water and brought home twenty mixed flounder, porgy, and hake that you plan to eat; you have twenty "fish" at hand. But if a scientist described how many different species made up that catch, they would say three "fishes"; the word *fishes* describes diversity.

Just a dozen or so diadromous fishes occur on the East Coast of North America. Only one is catadromous: the American eel. The best-known anadromous species of this region are Atlantic salmon, striped bass, American shad, alewife, and Atlantic sturgeon; others include sea lamprey, shortnose sturgeon, blueback herring, hickory shad, rainbow smelt, Atlantic tomcod, and Atlantic whitefish. Striped bass are also found along the Gulf of Mexico coast, together with the Alabama shad, skipjack herring, and a subspecies of Atlantic sturgeon.

But the boundaries among diadromous fishes and non-diadromous fishes are not well defined. Some view the Atlantic tomcod as anadromous, though they tend to migrate only as far as the saltier extremes of their home estuaries, and not into true marine waters. Then there are those that are "facultative," meaning that they can lead an anadromous life-history pattern if they want to, but it's not obligatory. Native brook trout and nonnative brown trout both may drop down rivers and go to sea, but most don't; the majority of striped bass in mid-Atlantic rivers migrate to the ocean, but some never leave fresh waters. Finally, almost all of these species may "landlock," completing their life cycles in fresh waters, such as lakes and reservoirs.

To view these fishes as a group is to be struck by their diversity of size and shape, a clue that there is little evolutionary commonality among them. "Primitive" sea lamprey have no bones and are jawless. Sturgeons also are considered primitive, have only cartilaginous skeletons, but display armored plates of bone on their flanks. Salmon and shads are intermediate on the evolutionary tree, lacking fin spines, with spiny-rayed striped bass being the most "advanced" among the entire East Coast lot. It's important, though, to not misunderstand the significance of loaded-sounding terms like *primitive* and *advanced*. *Primitive* is not a value judgment, and it does not mean such species are slow and dim-witted; lamprey and sturgeons would not have survived for hundreds of millions of years if that were the case. This term means only that these highly successful groups appeared early in fish evolution; *advanced* means they evolved relatively recently.

This begs the question: If diadromy is rare, and is a characteristic that is spread across the evolutionary spectrum of fishes, under what conditions did it evolve? And are diadromous fishes sea creatures that adapted to freshwater, or vice versa? Here our understanding gets shakier. It's commonly accepted that in temperate latitudes, fresh waters are "safer" than the sea as places to leave eggs, to hatch, and to live as larvae and juveniles. Marine waters undoubtedly are more ecologically productive, but beware—there are many more hungry mouths and sharp teeth there.

A population of an anadromous species can claim the benefits of higher survival for its young by placing them in freshwater where they rear to a size large enough to enter the sea to grow further and mature—but this strategy does not come without costs. The spawning adults need to expend considerable energy migrating to a suitable river, make the physiological adjustment to freshwater, expend further energy moving upriver against the current, in many cases feed little or not at all while in the river, expose themselves to river-based predators, complete the act of spawning and then migrate downstream, undergo another physiological adjustment, and then migrate back to their marine feeding grounds. That this grand but highly demanding life cycle is worth the costs is clear from the plenitudes that once existed for unadulterated populations, but the scope and complexity of this strategy leaves them especially vulnerable to the hand of man.

Interestingly, whereas anadromy dominates temperate latitudes, catadromy is more often seen in the tropics. The relative productivity of fresh and marine waters switches there; in these warm regions, fresh waters offer richer food webs than does the sea, and many species drop down rivers to the salt to spawn, with the young later penetrating inland. This is taken to an extreme with the Hawaiian gobies, marine fish that evolved to climb torrential rivers and waterfalls soon after transforming from larvae in seawater, this after being spawned at high elevations and washing downriver. The gobies' pelvic fins have transformed to sucking discs that allow them to inch along wet rocks, and the mouth of one species actually moves in only thirty-six hours from the fish's front end as a larva to the bottom of its head as a juvenile, to serve as an extra sucking disc. Some of these gobies were found above a 1,148-foot-tall waterfall, a feat of climbing more than five thousand times its body length. In human terms that's like scaling Mount Everest.

If gobies win the award for mountain climbing, eels are the winners for epic distances. My personal epiphany about the intense migratory drive of eels occurred in Iran. I was being shown around a sturgeon hatchery on the shore of the Caspian Sea. On one wall was a poster of the "Fishes of the Caspian," which included the European eel, a species that spawns a hemisphere away in the middle of the Atlantic Ocean. I asked, "How is this possible? The Caspian Sea is landlocked and sits one hundred feet below ocean level." The answer amazed me. Some eels, born in the Sargasso, loop around the Gulf Stream for 300 days, swim though the Straits of Gibraltar, down the Mediterranean past Sicily, past Istanbul and through the Bosporus Strait, across most of the Black Sea, and through the Strait of Kerch, before entering the Don River and locking through to the Volga, at long last gaining entry to the Caspian to the south, or farther upriver toward Moscow. Altogether, about a 4,000-mile trip—one way.

So how did the rare but supremely successful life-history mode of diadromy evolve? Mart Gross, a conservation biologist at the University of Toronto, hypothesized that in the case of anadromy, there is an in-between stage called *amphidromy,* much like for the tomcod mentioned above. In one scenario, a freshwater species makes occasional forays into brackish waters to feed, realizes some benefits (such as increased food availability), and this positive selection results first in amphidromy, where a population ventures to higher-salinity estuarine waters as part of its life cycle, with this adaptation eventually leading to anadromy, with migrations to the sea. And vice versa for catadromous fishes.

These concepts are appealing and make sense. But are they true? The late Robert McDowall, a prominent New Zealand fish biologist, thought not. When he looked at the evolutionary trees within taxonomic groupings of fishes, he found a mixed signal—that anadromy likely arose from both freshwater and marine origins. He also found no evidence for amphidromy being a precursor to catadromy. Indeed, McDowall believed that for some fish, diadromy may even be the ancestral condition.

Regardless of how it evolved, it is common knowledge that salmon "home" to the river they were born in. In fact, it's known or safe to assume that each of the Atlantic anadromous fishes homes, with one major exception: sea lamprey. What does it mean, to "home"? And to "stray"? Why should homing even occur?

Homing means that after spending months to years at sea, maturing, possibly far from its natal river, there is an overwhelming propensity that an individual will return to spawn in that same river. How these fish navigate in the sea to find the river they were born in is not completely understood, but research with salmon has shown that they "imprint" on the odor of their natal system before they go to sea, and that they detect that odor once again as they approach their river from the seaward side. Homing over the short term has been studied directly by tagging fish and seeing how many turn up again in their natal rivers, versus other rivers. (However, scientists have discovered that attaching tags to fish can alter their behavior.) Homing also has been studied indirectly using genetics; the more different two populations are genetically, the less gene flow between them is indicated, meaning that straying is therefore rare. This approach provides a long-term signal but not much present-day information. Either way, though, homing rates of about 98 to 99 percent or more seem to be the norm.

What then of the 1 percent, give or take a little, that end up spawning in a river other than the one they hatched in? This might seem maladaptive, but it's not. Consider perfect homing. If a river's population went extinct, it would *never* be recolonized—there would be no source of new individuals. Nor would colonization occur in any new, suddenly accessible habitat.

But the tendency to come back to the same river for generation after generation does have consequences—good ones from the fish's point of view. This is the engine for the exquisite fine-tuning of anadromous fish to their own life-history circuit. Different stocks of a single species, say, shad or salmon, mix in the sea, yet there may be noticeable differences among them in various characteristics, telling us that these differences are driven primarily by their particular freshwater conditions. How so?

The answer is that the fish become physically sculpted to the unique demands and opportunities provided by their fresh waters. Anadromous fishes once displayed remarkably recognizable variation below the species level, tuning expressed strongly enough to constitute a variety of groups that were seasonally or geographically sufficiently different to be termed "substocks," "races," or "runs," with many given colloquial names. Commercial shad fishermen in the Hudson were particularly attuned to variations, recognizing yellowback, blueback, greenback, golden, pink, pink-faced, locust, chunker (exceptionally deep-bodied), chunk head, and red-finned (possibly due to slight damage to capillaries in the skin) shad. Farther south in the Potomac and North Carolina, fishermen in the 1800s noted "May shad" late in the run that were fatter and deeper-bodied, with a thicker tail section than the earlier fish.

Atlantic salmon are especially adaptable in their physical and life-history characteristics, and the differences that emerge can often be linked to their migratory challenges. The salmon of the Grand Cascapedia in Quebec are large and powerful, reflecting qualities of that river. The Sevogle, a small branch of the Northwest Miramichi in New Brunswick, has small but very stocky fish. The Serpentine River, a tributary of the Tobique, also in New Brunswick, produces strong, wiry fish from its shallow, rocky stream. Maine's greatest river, the Penobscot, has good-size, muscular salmon. From Scotland, the Tweed produces bulky salmon. But salmon from Scottish Highland rivers, such as the Dee, with its upstream rapids, are lean but nicely proportioned. New Brunswick's Restigouche people adopted the salmon as their tribal symbol, adorning their canoes, clothing, and bodies with images of the fish. So intimate were they with salmon that it was said they could immediately identify which river a fish came from.

Salmon in long, fast rivers such as the Alta and Vosso in Norway require a large amount of energy to reach the spawning grounds, but larger rivers also are more likely to have enough water each year to support reproduction, so these rivers will select for a longer period of feeding at sea, and hence, for delayed breeding, with less repeat reproductions. Salmon in short, more easily traversed rivers with more uneven flows are typified by reproduction at an earlier age, but with more repeat spawnings, they are more apt to bet-hedge by spreading reproductive risks across years (e.g., the small salmon of the little "spate" rivers of Cape Breton, Nova Scotia). Most Atlantic salmon rivers worldwide also have some proportion of "grilse," which are individuals that spend only one winter at sea. Though quite small and composed mostly of males, they do help assure that some portion of the population returns to continue it.

Just how fine do anadromous fish take this fine-tuning? Late-run spawning salmon on the Miramichi enjoy post-spawning survival rates substantially higher than those of early-run counterparts, with early-run fish pushing into the headwaters and late-run fish spawning farther downstream. Within all the sections of only one large drainage, New Brunswick's Saint

John River, researchers found as much variation in reproductive character-istics of shad as found among all East Coast populations. For anadromous fishes there also are energetic demands on how often an individual spawns in its lifetime. The cost in energy of migration, plus the act of spawning, is about 60 percent of that stored in Atlantic salmon. For American shad it is as much as 70 to 80 percent in Florida's warmish St. Johns River, where there are no repeat spawners. In a northern river such as the Connecticut, where shad may spawn multiple times, the cost in energy on a spawning run is 35 to 60 percent. Nature's knife puts the slice between life histories where a fish spawns once instead of twice or more within the 60 to 70 percent energy-depletion range. Darwin would not have found these adjustments as dramatic as those seen in his Galapagos finches, but natural selection works on river fishes just the same.

Fortunately for their management, the effects of the particular environ-ments on these fish that home—and thus build up slight but important differences through natural selection—can be used to identify the popula-tion of origin where they mix in the wild. Striped bass, in particular, have received enormous attention toward discriminating between individuals from the Hudson versus the Chesapeake, and sometimes North Carolina's Roanoke River, too. In a sense, this science of stock identification has taken these fish apart, looking for useful differences.

In 1989 I drove some five thousand miles on coastal highways collecting about five hundred striped bass for a study, to look for differences among populations using the same specimens by different researchers employ-ing their own approaches—genetics, body shape, scale and fin-ray counts, scale shape, and fatty acids—allowing me a unique opportunity to see many fish from different rivers during the same season. Though I could not have assigned individuals with certainty to their rivers of origin like the Restigouche with their salmon, some generalized differences were vis-ible to my naked eye: Hudson River specimens were a distinctly mixed lot, Roanoke River stripers were compact, Choptank River fish seemed like classic "textbook" stripers, but the ones from the Rappahannock were long and sleek, like graceful athletes.

Gathering these specimens often meant meeting state biologists and helping to net the waters with them or picking up at the dock fish already caught. Either way, though, I needed to process and preserve the critical portions of the fish for the researchers. This meant creating "laboratories" on the fly. One time I obtained about thirty large stripers from the Choptank River and then rented a motel room in southern Maryland, spreading them over every horizontal surface and working them up. I sometimes wonder what the motel owners thought went on in that room when the next morning they discovered a guy with New York license plates on his van had left trash cans overflowing with bloody newspapers and dozens of syringes.

The constant washing of a watershed with rainfall and snowmelt slowly depletes the nutrients that sustain its ecological productivity. But the relentless circularity of the anadromous life-history cycle helps return some minerals from richer marine waters back to rivers. Anadromous fish themselves are bundles of nutrients—that's why we eat them. Once having left their natal rivers as young individuals just large enough to have a chance to survive in the richer but also more dangerous sea waters, they feed heavily and put on weight, eventually maturing and becoming egg- or milt-laden and ready to spawn. And so a river trades numerous young sent to sea fueled by river-derived nutrients for fewer but much larger adults that are themselves laden with marine-derived nutrients upon their return.

These migrating spawners bleed some of these chemical compounds to a river as they excrete waste products. More are contributed in the many eggs and sperm cells that don't find partners or that perish after fertilization. But the largest nutrient inputs originate from the adults that die in the river, more often during the post-spawning phase of life. For some anadromous fishes death soon after spawning is programmed into their genes. For Pacific salmon, in which decomposition seems to precede death, nutritious hunks of salmon are so commonly seen in the flow during spawning runs that fly fishermen use "flesh flies"—feathers tied to resemble ragged pieces of salmon—to draw strikes from the salmon that have yet to spawn.

The contribution of dead salmon to the fertility of Pacific rivers cannot be overstated. Qualitatively, it seems obvious. Visit the spawning reaches of an unadulterated stream during a run and carcasses lie in the water and on the banks in various stages of rot as still more fresh bodies beat their way monomaniacally past them, only hours to days behind but in lockstep. Juvenile salmon already can be seen nibbling on the bare flesh of their deceased relatives, part of a suite of insects, fish, birds, and mammals that will scavenge them.

Quantitatively, their importance ripples through the Pacific slope ecosystems. Ninety percent of a Pacific salmon's weight is gained at sea. In *King of Fish,* David Montgomery writes: "Up to a third of the nitrogen in valley-bottom forests swam up the river as a fish." Trees growing along salmon-bearing streams grow up to three times faster than those living along salmon-free streams. Higher in the food chain, more than 90 percent of the nitrogen contained in Alaskan brown bears comes from salmon. Circularity—relentless circularity. The hordes of salmon smolts sent seaward could never reach both the abundances and sizes without the lagged enrichment provided by their parents. Salmon essentially extend the fertility of the oceans inland for their own purposes, but also to the benefit of a host of other species tightly entwined in these special ecosystems.

Atlantic salmon did not evolve with that same death switch. And New England and Canadian forests along salmon rivers—as verdant as they are—do not display the grandeur of their cross-continent counterparts. No one has really satisfactorily answered why the salmons of two ocean basins don't share the same life cycle. But because a phenomenon doesn't reach an extreme doesn't mean that it's unimportant. In fact, for Atlantic salmon, surviving first spawning and then returning appears to be the exception; a rough rule of thumb is that one in ten comes back to spawn a second time.

Other East Coast anadromous fish contribute essential nutrients to rivers too. Phosphorus is usually the limiting element in fresh waters. A dead adult alewife adds more than one-half a gram of phosphorus to the ecosystem, while a spawner that survives excretes about one-third of that amount.

This may not be much on an individual basis, but pristine runs that numbered in the tens to hundreds of thousands to millions would have mightily enriched the river ecosystems they spawned in. In fact, for one small Connecticut pond, Yale researchers estimated that at moderate abundances, more than 40 percent of the phosphorus found there arrived in the form of alewives. Likewise, in a modest stream in Massachusetts, sea lamprey, which always die after spawning, were found to add about a fifth of all the phosphorus that entered that reach annually. In these and other Eastern Seaboard rivers, anadromous fishes when they still ran silver did much to extend the influence of the Atlantic inland. But Atlantic rivers are different today; ghosts aren't corporeal, and ghosts don't migrate.

Chapter 3

The Seasonal Parade

There is no measure in the world of nature more excellent than a fish.

—*John Hay,* The Run *(1959)*

How best to introduce the great freshwater-sea migratory fishes of the Atlantic? There is a natural order that has occurred for millennia; I will let them swim into your consciousness in the same annual sequence they would materialize from the ocean to drive their way up a free-flowing river.

Ocean-River Migrants

February
Rainbow Smelt *(Osmerus mordax)*

Rainbow smelt will never be counted among the world's great gamefish. Yet they do have a unique and delicate flavor reminiscent of violets and cucumbers, and they offer welcome angling during cold months when few other fishes are available. Smelt also contain healthy oils; a close Pacific relative, the eulachon, is so fatty that Native Americans would dry them and then light them as torches. A boreal fish that visits north temperate rivers, these days rainbow smelt reach as far south as Massachusetts in catchable numbers. They also make the earliest spawning migrations of all the Atlantic anadromous fishes, sometimes entering rivers before ice-out. Smelt don't make dramatic runs far inland, but they do usually spawn somewhere above the head of tide.

Rainbow (Eastern) smelt
NOAA/Department of Commerce

A smelt is not a big fish; a twelve-incher is a trophy. Nonetheless, a smelt is an effective predator within its own realm, sporting a set of strong canines in an angular jaw—ready to snap up shrimp, other small crustaceans, and little fish—and propelled by a sleek and silver-striped trunk. As diminutive as anadromous smelt may be, they even have populations of dwarfs—mini-smelt that don't exceed about five inches—that coexist with normal-size smelt in some freshwater lakes.

Smelt are a zoogeographic anomaly. The smelt that range the Northwest Atlantic today lost access to all the rivers covered by glaciers during the Pleistocene, an epoch that ended only recently in geological time—on the order of 10,000 to 15,000 years ago, depending on latitude. How did smelt survive? DNA analysis showed two major groupings that coincided with two glacial refuges, termed *refugia*. Some smelt populations, called the "Atlantic" race, probably were pushed down the coast, to unglaciated rivers on the coastal plain from New Jersey on south. Others, termed the "Acadian" race, rode out this period, spawning in newly formed rivers on the offshore Grand Banks that rose above the ocean's surface as sea levels fell worldwide. Later, the two races came into contact in the St. Lawrence estuary—the Acadians from the mouth, and the Atlantics via a temporary freshwater route through the Champlain Valley. DNA findings showed that even though some intermixing was detected, individual runs of smelt tended to be made up of only one race, suggesting that some reproductive isolating mechanisms developed when the two groups were separated; perhaps an early stage of speciation?

19

Little is known of smelt at large in the sea, but their coming inshore in late autumn and winter prior to spawning once made them great democratizers. In the Boston area in earlier times, smelt crossed all social barriers, with businessmen joining the hoi polloi angling in the early hours of late-autumn mornings for breakfasts of smelt. Anglers still fish for them off docks at tidal rivers near Boston and other New England locations, but not to the south, where they have largely disappeared.

For truly dedicated smelters there are ice fisheries in rivers, particularly in Maine and Maritime Canada. In Maine you can rent a shack on a frozen river for a day or night, complete with stove, wood, electricity, and chairs. Inside the ice shanty, as it is called, is a horizontal bar from which are attached about six to twelve baited lines that are suspended from floats on the open water of the hole that's been cut. See a bobber twitch and haul up a smelt. In between bites, smelters drink coffee or spirits and share stories; if they do well, they may even fry up a few smelt right there on the woodstove. At the end of a session, a good catch may be measured not in counts of individual fish but in numbers of quart containers.

March
River Herrings or Alewife *(Alosa pseudoharengus)* and Blueback Herring *(Alosa aestivalis)*

Come early spring, a few communities along the Eastern Seaboard still are energized by the knowledge that the herring are back, once again penetrating every stream or creek of any consequence. Some rush to the waters to gather them for bait, others for a meal, and many just for the psychic refreshment of this seasonal signifier of the renewal of life. Peer from a bridge or other vantage point at the height of a run and you may see thousands of determined silver bodies beating their way against the flow, sometimes leaping and climbing through chutes in rapids and small waterfalls.

The unacknowledged poet laureate of anadromous fish, John Hay, did not rhapsodize about leaping salmon, brutish sturgeons, or predatory striped bass; instead, he honored the little alewife, his insights serving as

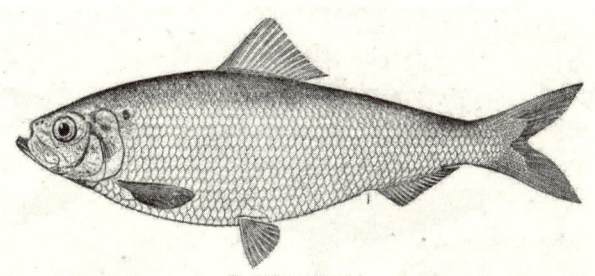

Alewife
NOAA/Department of Commerce

proxy for the commonalities among all of these remarkable migrators. He did this in a little 1959 masterpiece of revelation and understatement titled *The Run*. John Hay was a Harvard graduate, reporter, and poet, who later directed a children's museum and lived on a hilltop overlooking Cape Cod Bay, his station for chronicling the importance of these little fish in the lives of the local populace. *The Run* includes many marvelous observations, among them this essence of the species: "Fragile they are, and powerful, a wonderful work of which so many are made as to afford them death as well as life."

River herring is a term of convenience for these two closely related species. It's not clear how the intriguingly named alewife gained that appellation. As for any thought-provoking name, the etymology is muddled, with theories including derivation from an Indian term for the fish, a French name for a similar species, and for its deep belly, reminiscent of an old-time female tavern keeper. All over Atlantic Canada, river herring are called by the Acadian name, *gaspereau,* possibly after an eponymous Nova Scotian lake, or vice versa. A commonly used name in the nineteenth century was "branch herring," whereas the slightly slimmer blueback herring was known as "glut herring."

Where they co-occur alewives run earlier, but overlap with bluebacks, the alewives often entering headwater ponds and the bluebacks preferring to spawn in moving water. And they are cut from molds similar enough that many have trouble distinguishing them. (One sure way is to check the color of the abdominal cavity lining; it's pale in alewives, and dusky black in blue-backs.) Alewives are also more northern in their distribution, maintaining populations between Newfoundland and North Carolina; blueback herring stocks are found between New Brunswick and Florida. Both species readily landlock, and alewives, in particular, now occur in many lakes and reservoirs, often in reduced sizes.

Though important as a food source historically, the few river herring consumed today are first smoked. But there is increasing recognition of their critical role as prey fish for larger marine species and as vital ecological links between smaller streams and the sea; that is, they are among the small fishes that power marine food chains, as what Paul Greenberg in *Four Fish* called "the silver coin of the realm." In the ocean, river herring eat plankton and fish eggs and larvae, often mixing with the true sea herring, which does not enter freshwater. In fact, *pseudoharengus* means "false herring."

The ability of river herring to pick their way through crashing high-gradient flows makes them uniquely suited to climb fish ladders. On Cape Cod there is a tradition of "herring wardens" who monitor and protect individual runs, which may involve installing a seasonal ladder at a dam or keeping a permanent ladder free of debris. An especially venerated fish ladder can be seen in Damariscotta Mills, Maine, where it forms the basis for two annual festivals. This mostly naturalized ladder of cement and rock pools was constructed in 1807 to bypass a sawmill dam built in 1726. Today residents and alewife aficionados from afar gather to celebrate the return in spring of some half-million adult fish to this little system, marking it with music, "chowdah," science exhibits, and even an alewife puppet theater and a chance to be photographed being pretend-lifted like an alewife to the sky by a giant osprey. They do it again in the fall, which includes the "Running of the Alewives" race, where children compete to have their wooden hand-painted alewives drift down the ladder, symbolizing the completion of the species freshwater life-history loop, mixed with legions of real juveniles dropping down to the sea, as a silvery mass of fragility and power.

April
American Shad *(Alosa sapidissima)*

If there is a quintessential and broadly important East Coast anadromous fish, it is the American shad. In early colonial times shad poured by the millions into dozens of rivers from Quebec to Florida. And not just into them, but before damming, far up them, deep into the coastal plain to spawn in the fast waters of the Piedmont or to reach headwater lakes. Almost 125 miles up New Hampshire's Merrimack to Lake Winnipesaukee, 500 miles up the Susquehanna to Lake Otsego in New York, 300 miles up Georgia's Altamaha, and 375 miles up Florida's St. Johns. Runs that ran silver up and back down in some rivers would last for months. Moreover, the fish were large bundles of protein, with many weighing four or five pounds, and some exceeding ten pounds.

The oily meat of shad was appreciated by Native Americans and colonists, sometimes being called the "poor man's salmon." Indeed, *Alosa sapidissima* translates as "most delicious of herrings." But eating shad comes at a cost in convenience, since a shad has more than 700 mostly fine bones. So

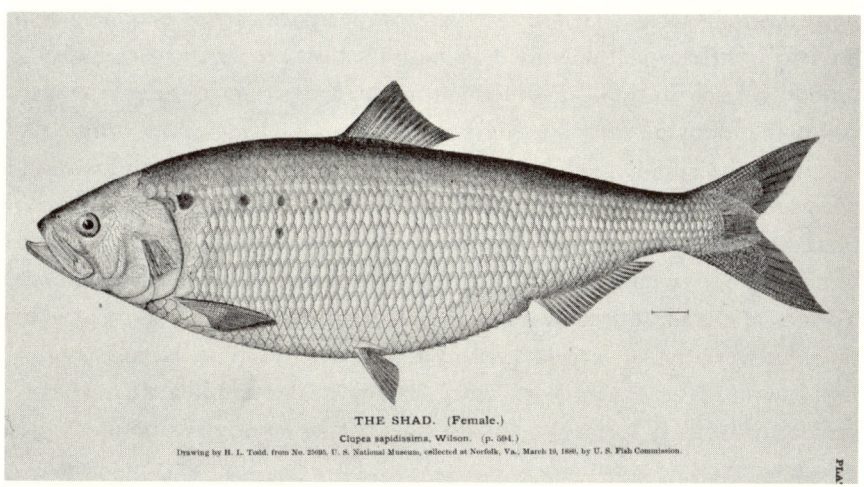

THE SHAD. (Female.)
Clupea sapidissima, Wilson. (p. 594.)
Drawing by H. L. Todd, from No. 25695, U. S. National Museum, collected at Norfolk, Va., March 19, 1880, by U. S. Fish Commission.

American shad
NOAA/Department of Commerce

bristly is the internal anatomy of a shad that a Micmac Indian legend tells of an unhappy porcupine asking the Great Spirit to turn it into some finer form of animal. The spirit grabbed the porcupine, turned it inside-out, and tossed it into the water to become the bony (but still delicious) shad.

Because it wasn't always convenient to come to the fish, shad peddlers made a living bringing the fish door-to-door to the people. This continued for a century in towns along the Susquehanna, the fishmongers announcing their arrival by blowing on a special horn and then crying "shad-o-o-o-oh." Today shad are also known for their roe, which has been lauded as the foie gras of seafood. Shad roe was not always esteemed; until the mid-1800s much of it was fed to pigs and given to the poor. But by the turn of the twentieth century, appreciation had grown, with special knives being invented to extract the egg masses without splitting the delicate skein sac. Eating shad roe became a cultural rite of spring. As did planked shad, a process where boned shad were hung on spikes on oak boards, sometimes laced with strips of bacon, and then slowly roasted as they were leaned near hot coals. Not only was planked shad popular in restaurants, but planking also became a central feature of shad festivals, events held on many rivers to celebrate the return of shad as a marker of spring and celebration of providence.

John McPhee chronicled the shad's importance in *The Founding Fish,* the book's title a well-defended opinion on his part. Shad were indeed a founding fish, if their involvement with American presidents counts. Thomas Jefferson, who was born at eponymous Shadwell, Virginia, just east of Monticello, hauled seine for shad in his younger years and was said to love a fine shad roe soufflé. George Washington, after returning from the French and Indian Wars and resigning his army commission, devoted himself at age twenty-six to living on and running his plantation, Mount Vernon, about a hundred miles up the Potomac from Chesapeake Bay. This included activities such as overseeing his slaves, growing tobacco, fox hunting, attending theater and cockfights, playing cards and billiards, and a task he took extremely seriously, harvesting shad—two decades before Valley Forge where shad were said to have saved the General's starving troops.

Shad young spend their first months slowly working their way down their natal rivers, entering the sea in autumn to begin annual migrations in large schools along the continental shelf, sometimes as deep as seventy fathoms. There they fatten mainly on zooplankton and small fish. But with such a wide latitudinal range of populations, these coastal migrations occur as distinct loops in the southern and northern portions of their range. Shad are thought not to feed during their spawning migrations in freshwater, and they can be challenging to entice on hook and line, yet some do strike—a mystery best explained by the fact that there is almost nothing actually available of their usual food items in rivers in springtime.

I once attended a gourmet dinner in Manhattan in which shad was served as a novelty in several creative courses, with speeches made between them on the lore and romance of the fish. What a far cry from earlier times when they were the daily grist, and even a form of currency. For the Susquehanna fishery near Wilkes-Barre, McPhee wrote, "The fishermen did their fishing after dark. They drank 'old rye.' Customers bartered with them, paying whiskey for shad. They were also paid with leather, iron, cider, maple sugar ('one good shad was worth a pound of sugar') and cider royal (cider and whiskey). A bushel of salt bought a hundred shad. Walter Green, of Black Walnut Bottom, 'gave twenty barrels of shad for a good Durham cow.'" Nowadays it's tough to find a shad to trade for, even with a barrel of cash.

May
Striped Bass *(Morone saxatilis)*

The striped bass, much like the Atlantic salmon, is a noble species; the difference is that it is eminently accessible to the common person. Striped bass, known as the "striper" from New Jersey northward, and rockfish (or "rock") to the south, is a river-born fish that nonetheless rules inshore coastal waters. It is at home in a remarkable variety of habitats, from breaking waves in the sandy surf, to rocky, current-washed reefs, to the muddy shallows of salt marshes, and to fast-flowing rivers. It is also the consummate urban species, flourishing in metropolitan harbors such as in Boston, New York, and Baltimore, where it ambushes baitfish prey from behind man-made pilings, or at night at the shadow lines of bridges.

Striped bass
NOAA/Department of Commerce

Striped bass have different migratory natures depending on their latitude. The truly large, ocean-going populations are spawned in the Hudson
and Delaware Rivers, and the many tributaries of Chesapeake Bay. South
of Virginia all the way to Florida's St. Johns River, and again, in rivers
along the Gulf of Mexico coast, these nonmigratory striper populations
are orders of magnitude smaller. From New England to the St. Lawrence
River, Quebec, striped bass may be locally migratory, but their populations
are no more than modest and are poorly studied. In most rivers striped bass
deposit their neutrally buoyant spawn upriver of the "salt front," where
brackish water is last detected, though the population in North Carolina's
Roanoke River instead uses powerful rapids to keep its eggs drifting till
they hatch.

Striped bass are esteemed by many for their moist, flaky meat. Because
they occur in so many environments, they've been harvested since colonial
times with all sorts of commercial gear, from haul seines to gill nets to trap
nets to trawls to rod and reel, and more. Early on, in 1634, William Wood
in New England's Prospect gave these instructions: "The way to catch them
is with hook and line, the fisherman taking a great cod line to which he fas-
teneth a peece of lobster and threwes it into the sea. The fish biting at it, he
pulls her to him and knocks her on the head with a sticke." And even more
resourcefully, in the 1800s they were hunted on Chesapeake Bay's Eastern
Shore on horseback, with spears. One account: "The large fish coming to

feed on the creek shores, overflowed by the tide, showed themselves in the shallow water by a ripple before them. They were ridden on behind and forced into water too shallow for them to swim well, and were speared." Later in the nineteenth century, sportsmen built rickety "bass stands" of iron and wood on the rocky ocean shores of Massachusetts and New York, again fishing and chumming with lobster (then highly abundant and poorly regarded as human food).

Lobster is only one item in the highly varied diet of striped bass. Any smaller fish that can be engulfed, will be, by the suction it creates when it flares its gills, but stripers have a special fondness for rich, oily fishes such as eels, mackerel, and menhaden. They also inhale squid, crabs, shrimp, sea worms, and even clams washed out of the bottom by storms. In the urban waters of the Hudson, some fishermen catch them on slices of beef liver.

It is the dream of every striped bass angler to land a "fifty," a fifty-pounder (or larger). But stripers of every size class provide sport and do much to support the New England and Mid-Atlantic recreational fishing industry, from small "schoolies" to huge "cows." (All truly large striped bass are female; males rarely exceed twenty-five pounds.) Entire communities, such as Provincetown, Massachusetts; Montauk, New York; and Cape Hatteras may be overrun with anglers when a solid striped bass run develops, typically, in autumn. In springtime, migratory stripers move northward to rich summer feeding grounds off eastern Long Island, Cape Cod, and Maine. But in autumn all hell breaks loose as schools of stripers fattening on their southward migrations attack schools of baitfish, often showing on the surface as acres of splashes marked by clouds of seabirds looking for an easy meal of the frightened prey—a behavior called a "blitz," after the German term for all-out war, *blitzkrieg*.

But unlike the Atlantic salmon, a species reserved largely for the wealthy, striped bass, being found near shore along much of the Eastern Seaboard, are available to anyone with a decent rod and reel and the patience to master the knowledge to outwit a truly challenging gamefish. Anglers from every walk of life become passionate—even fixated on stripers. Some chase them relentlessly from boats, join clubs, and craft their own lures.

But no fishermen are more hard-core than the striped bass surf caster. Peter Matthiessen described the obsession: "The unseen quarry and mysterious dark water, the pleasure taken in the strong and skillful cast, the sound and smell of sea and weather, the healing solitude, and the suspense, are reward enough for the true sportsman who seeks no profit from the hobby, and surfcasting for striped bass probably claims more fanatics than any other form of saltwater fishing."

June
Sea Lamprey *(Petromyzon marinus)*

The sea lamprey is the black sheep among the anadromous fishes of the Eastern Seaboard. But like so many outcasts, it is also profoundly misunderstood. This primitive, eel-like species conjures all the horror to be expected of an oversize parasite. Indeed, a sea lamprey recently attached itself to the back of a human swimmer in the English Channel (remaining there until the swimmer completed his crossing, after which the "host" had it promptly removed). A similar experience happened to a swimmer making a record crossing of Lake Ontario (sea lamprey also occur in the Great Lakes). She felt a gnawing sensation at her middle—a sea lamprey had bored right through her bathing suit. "I struck hard at it with my hand and my blow knocked it off," she said, allowing her to complete the twenty-one-hour swim.

THE SEA LAMPREY.
Petromyzon marinus, L. (p. 677.)
Drawing by H. L. Todd, from No. 10654, U. S. National Museum, collected at Wood's Holl, Massachusetts, by Vinal N. Edwards.

Sea lamprey
NOAA/Department of Commerce

Sea lamprey lack jaws; instead they have a bristly sucking disc, the center of which displays a cluster of forward-facing teeth. It's safe to say the average person does not find this an endearing sight. Nor are sea lamprey something that other fish wish to find anywhere near them. A sea lamprey can attach itself to a host fish, rasp a hole in its body, and suck out enough blood and body fluids to kill its prey while riding it for days, or even weeks, depending on its size. A secretion by the lamprey even stops blood from clotting. Not all attacks are fatal; the lamprey may let go of its host before it dies, allowing the possibility of recovery, or the sapped victim sinks to the bottom while the lamprey looks for its next meal. This stage of the sea lamprey's life cycle can last for two years, during which it may kill forty or more pounds of fish.

But before it becomes parasitic, the sea lamprey leads a surprisingly benign, homebody existence. It begins life in a rock rubble nest constructed by its parents in a cool-flowing river. The rubble must be of a certain size, and the flow must be of a certain velocity. Although greenish-brown in the parasitic phase, spawning lampreys turn a mottled orange-brown that makes them easy to spot in the shallows, like colorful two-foot-long hawsers. When sighted, they appear industrious, carrying rocks using their sucking discs to engineer the pebbly nest that rises off the bottom in an arc to steer the current to the developing eggs. Thoreau ably captured their grim determination: "They ascend falls by clinging to the stones, which may sometimes be raised, by lifting the fish by the tail. As they are not seen on their way down the streams, it is thought by fishermen that they never return, but waste away and die, clinging to rocks and stumps of trees for an indefinite period; a tragic feature in the scenery of the river bottoms worthy to be remembered with Shakespeare's description of the sea-floor." Thoreau was correct; the adults *do* die after spawning, contributing their nutrients to the watershed, and likely helping to nourish their young, as happens for Pacific salmon on a much grander scale.

After hatching, the sea lamprey young make an abrupt change in habitat, seeking silty backwaters where they lead lives not much different than clams. Young lampreys live vertically in burrows, heads facing up, filter feeding. Take a backpack electroshocker—an electrical device scientists use to stun fish in order to collect them—and send the electrical current into the bottom of a muddy slough on a lamprey river. Out of the water will shoot little replicas of the adults, sometimes in great numbers. This burrowing phase can last up to about seven years, depending on food resources and growth rate, and near its end, the lamprey develops its adult parasitic armaments and heads downstream to the sea to begin its search for unlucky hosts.

One surprising thing Thoreau didn't know, however—or anyone else, for that matter, until recently—was that sea lamprey are the only Atlantic anadromous fish that doesn't home. A study I'd read had shown a complete absence of homing in a well-run experiment in landlocked Lake Huron, a result which jarred against everything I knew about anadromous fish. In a rare eureka moment, it occurred to me that the combination of anadromy and parasitism doesn't concur well with homing. Why? Unlike for other anadromous fish, which can migrate in a cohesive circuit in marine waters and easily cycle back to their home river, sea lamprey spend days or weeks at a time being dragged in all directions by a variety of hosts, leaving those that are born in any single river dispersed all over the ocean when it's time to begin their spawning migrations. Thus, it appears they utilize the nearest suitable river for their own reproduction, rather than the river they were spawned in. Not only did a signal of non-homing appear unequivocal when colleagues and I studied it using genetics, but the "nearest suitable river" theory is in keeping with a sophisticated chemical signaling between those clam-like young upstream and ripe adults downstream, which informs spawners that that river supports the species.

Being a cartilage-only, early evolutionary form of fish that predates jawed fishes doesn't mean lamprey don't have some admirable animal traits. (A lamprey fossil was recently found in a Cretaceous deposit that also yielded feathered dinosaurs, although some are older, from the Devonian.) Before I became familiar with them, I had a mistaken notion that they were doltish and slow-moving. To collect spawning lamprey is to tangle with a creature that thrashes angrily in response to the interruption. Once

I visited a fish elevator at the Holyoke Dam on the Connecticut River in Massachusetts. Peering down at a portion of the dam with only some leaky flows from above, I noticed lengthy fish leaping to move upstream. Salmon? No, they were sea lamprey, hurtling upward like spears. It seemed incongruous, but a sea lamprey must have "burst" swimming ability to close the gap and attach to a potential host fish that would certainly like to escape it. Sea lamprey also have one of the greatest (if not *the* greatest) vertical ranges of any non-avian vertebrate on Earth. A sea lamprey was trawled in the sea from a depth of 13,500 feet. Sea lamprey also run into tributary streams of Lake Superior, approximately 600 feet above sea level. Thus, not only does the sea lamprey have a vertical range of at least 14,000 feet, but, more remarkably, it also exists across two major realms, the deep sea and inland fresh waters.

If ever there was a species that engendered different responses by the public in different corners of the world, it is the sea lamprey. Sea lamprey is a delicacy in Europe—so much so that as for most kinds of fish with a high price on their heads, they are in short supply. Ancient Romans and upper classes of the Middle Ages esteemed lamprey; in fact, King Henry I of England is said to have died from eating "a surfeit of lampreys."

Along the US coast, sea lamprey are essentially ignored. Modest local fisheries existed in some communities that had first-generation European immigrants, but these have died away. The reaction to sea lamprey inland in the United States is virulently negative, and not without reason. The Great Lakes today are remarkably compromised ecologically from their precolonial condition. One major change is the presence of large numbers of sea lamprey, enough to drive a top-level fish predator, lake trout, to extinction in Lake Ontario, and to decimate lake trout stocks in the other four lakes. There is no doubt that sea lamprey invaded Lake Erie from Lake Ontario through the Welland Canal in 1921. But whether sea lamprey are native to Lake Ontario has fueled a major, long-running debate. Regardless, sea lamprey became so numerous that scientists could simply drag the bottom of the lakes with nets to pick up dead lake trout, their skin marked with lamprey scars. Vast sums have been spent by the United States and Canada for chemical weapons to fight lamprey, while, ironically, Europeans can't get enough of them.

July
Atlantic Sturgeon *(Acipenser oxyrinchus)*

Atlantic sturgeon are the dinosaurs of rivers. Armored with plated scutes along their flanks and with sucker-like mouths and Salvador Dali barbels, there is something jauntily primeval about them. Indeed, sturgeons do reach back deep in time, with fossils found dating as far back as the Upper Cretaceous, at least 100 million years ago. Sturgeons are *chondrosteans*—primarily cartilaginous, unscaled fish that dominated the aquatic realm for at least another 200 million years before that, to the early Jurassic, and possibly all the way to the Devonian, 400 million years ago.

Success for sturgeon is measured in the sheer longevity of their lineage, not in their diversity. Though they may have swum alongside true aquatic dinosaurs, there are only about twenty-four species of sturgeon in today's world (the uncertainty based on taxonomic questions and the conservation status of particular forms). North America has eight, plus one of the world's two paddlefishes, closely related filter feeders with broad, flattened snouts. The Eastern Seaboard is home to two of the continent's anadromous sturgeons, the Atlantic sturgeon being the behemoth of the pair.

The primitive appearance of a sturgeon is misleading. The sturgeon bauplan is supremely adapted to living on the bottom of large rivers and lakes and, for anadromous forms, the ocean. (This, of course, is tautological; nothing can survive on this planet for more than 1 million millennia without being adapted to its conditions.) But sturgeon have some physical and physiological surprises. Despite sporting a lengthy, flattened snout and a mouth that appears recessed, sturgeon somehow manage to consume their share of small fish, based on examinations of their stomach contents. They can use their snout like a blade to plow the bottom; in 1888 observers saw a school of Atlantic sturgeon work like hogs through the mud of Tampa Bay. Sturgeons also can "shoot" their mouth downward to capture worms, crabs, crayfish, and other invertebrates. Grasp the lower lip of a sturgeon and pull;

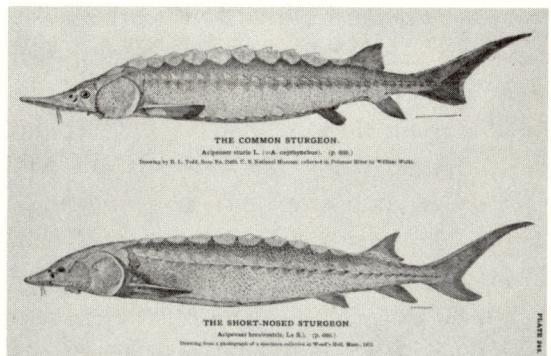

Atlantic sturgeon (with Shortnose sturgeon, below)
NOAA/Department of Commerce

you'll be amazed at how the mouth forms a delicate pair of lips at the end of a remarkably long tube. Sturgeon in rivers mostly live deep in murky waters where visibility is next to nil. But as they cruise, they are believed to detect food with still poorly understood electrical fields, in addition to those sensory barbels, before their mouths spring into action. Atlantic sturgeon also often show themselves by leaping. Many explanations for this have been offered, including that they are shaking off parasites, communicating with each other, and, most anthropomorphically, that it just feels good, but recent studies showed that it is a physiological need to obtain air to help their swimming stability.

Atlantic sturgeon once sustained populations in about thirty-five rivers between the St. Lawrence in Quebec and the St. Johns in Florida, a huge latitudinal range, from the semi-boreal to the semi-tropical. The number of rivers they inhabit has been reduced, and where they persist, their abundances have been radically lowered. Among the populations that remain reasonably robust are those of the St. Lawrence, Kennebec, Hudson, and Altamaha Rivers. But the river that once supported prodigious numbers of Atlantic sturgeon, the Delaware, is only now, after decimation in the "caviar craze" a century later, showing evidence that a few still spawn there.

The sturgeons' long duration on the planet corresponds with a sort of agelessness about them. My colleague David Secor at the Chesapeake

Biological Laboratory once described Atlantic sturgeon as a "pokey" fish; things take time with sturgeon. And a writer called sturgeons "a kind of philosopher among fishes, as if its ancient lineage had bred, over the thousands of centuries, a curious old wisdom and a quiet acceptance of change. The sturgeon has seen more years when it first spawns than many fish see in a lifetime." This protracted juvenile stage is pronounced in northern populations where cold waters lead to slower growth. Females in the St. Lawrence River may not become reproductively mature until age twenty-five, a remarkably late date for a fish, or any other vertebrate.

Spawning of Atlantic sturgeon takes place in late spring and early summer on rocky bottoms and reefs well inland in rivers. Young individuals as early as age two begin to leave their home rivers to launch an estuarine and marine existence; others linger for as much as several more years. During their period of extended adolescence, many appear to move north and south with the seasons from estuary to estuary, having no particular compunction to return to their natal river until their first spawning event. Once having spawned, though, females are in no hurry to do it again. The interlude is not fixed, running between two to four years. And they are the Methuselahs of the fish world, living as long as a half-century in northern rivers, but substantially less to the south.

Atlantic sturgeon, like all anadromous fishes, put on most of their weight in marine waters. This creates a challenge for another form of the Atlantic sturgeon that occurs in the Gulf of Mexico, known as the Gulf sturgeon. Nomenclature now gets confusing. The Latin name of the full species, Atlantic sturgeon, is *Acipenser oxyrinchus*. The Gulf sturgeon, *Acipenser oxyrinchus desotoi,* is a subspecies of the species known as Atlantic sturgeon. But the form found on the Eastern Seaboard, commonly known as Atlantic sturgeon, is also a subspecies, *Acipenser oxyrinchus oxyrinchus.*

Sophisticated DNA analyses indicate that Gulf sturgeon diverged from their East Coast relatives around the beginning of the Ice Age, 2 million years ago. The exact geological cause for this separation is not known, but may have involved the emergence of the Floridian Peninsula as waters became locked up in ice. Gulf sturgeon established populations from the Suwannee River in Florida westward to the Mississippi, but climate forced them to lead an odd marginal existence, so that their abundances never matched those on the Eastern Seaboard. The Gulf of Mexico is too warm for sturgeon for many months. So are much of the river waters they inhabit, but there are thermal refuges where cool waters vent from bottom fissures, allowing Gulf sturgeons a place to hunker down and ride out summer heat waves. There are ecological consequences to this defensive posture; the sturgeon schools quickly exhaust any local food resources and must deplete their energy reserves. When waters cool they run downriver to feed in the rich marine waters of the Gulf, but the net result is fewer and smaller individuals compared with their sister subspecies. Climatic warming will do no favors for Gulf sturgeon.

October
Atlantic Salmon *(Salmo salar)*

The second part of the Atlantic salmon's Latin name, *salar,* means "the leaper." Atlantic salmon are the acrobats of the Atlantic Ocean's anadromous fishes. Although rarely observed today because they are so scarce, when they were more abundant they could often be seen plowing through furious rapids and vaulting sizable waterfalls. This dramatic passage was driven by an instinct to deposit its eggs as far up watersheds as possible, where predators were few. Many anadromous fishes spawn in the main stems of large rivers, sturgeons on deep reefs, shad in big pools, but salmon go way, way beyond, driving their athletic bodies to small feeder streams where their considerable size seems out of proportion to the setting.

Unlike for the other anadromous fishes of the Eastern Seaboard, the earliest written mention of Atlantic salmon was not colonial, but Norse. Leif Ericson established the Vinland colony circa 995 at a site now known as L'Anse aux Meadows on the Newfoundland coast. The old saga of Eric the Red, which described Leif's adventures, states: "We found that there were quite a few salmon in the river, especially in the autumn, and sometimes we even caught one with our bare hands. The fishermen also from time to time made excellent catches of salmon out at sea and the fish were, on average, larger than their relatives we knew from Greenland."

THE ATLANTIC SALMON.
Salmo salar, L. (p. 468.)
Drawing by H. L. Todd, from specimen in the U. S. National Museum, taken in the Delaware River.

Atlantic salmon
NOAA/Department of Commerce

A synthesis of information on the original salmon rivers of New England lists 196. They ranged from the Aroostook River in northern Maine, a branch of the St. Croix that separates the United States and Canada, to Connecticut's Housatonic River that empties into Long Island Sound, just east of New York City. But this number masks the diversity of salmon rivers; some are just short unbranched flowages, whereas others are tributaries to major rivers that supported spawning salmon. Connecticut exhibits both kinds; the tiny Hammonasset is not far from the mouth of the Connecticut River, which, according to the list, once had an additional forty-seven branches that supported salmon.

Salmon spawn in autumn in gravel redds built by females who turn sideways and lash at the rubble with their powerful tails. Unlike species such as alewives that have highly defined spawning runs, North American salmon may enter rivers as early as March and then wait for fall. The fertilized eggs lay in the gravel all winter, protecting them from predators. Eggs then hatch the following spring, and the young "parr" lead a life much like stream trout for one to a half-dozen or more years, feeding mainly on insects. But as they gain in size, an important transformation occurs: They eventually *smoltify,* a suite of physiological, morphological, and behavioral changes to prepare for sea life. These five- to nine-inch smolts shed their camouflaging vertical stripes and become silvery, to better blend with seawater, hormonal changes gearing them for higher salinities and faster growth. The young salmon then enter marine waters where they feed on fish and zooplankton and mature, but where their deeper secrets are yielded only iota by iota. In the sea American salmon mingle with salmon from Canada, Greenland, Iceland, and many European rivers ranging from Russia to Spain. After this two- to three-year sojourn (for most individuals), they return to spawn, often as silvery eight- to fifteen-pounders. But unlike Pacific salmon, Atlantic salmon may spawn more than once, and some return to rivers as forty- and fifty-pounders.

As befits such an intensely appreciated fish, there is a lexicon to describe key life stages. As mentioned, *grilse* are individuals that return to spawn after only one "sea winter." *Kelts,* or black salmon, are individuals that have finished spawning and are heading downriver back to the sea; they are lean and hungry, and usually off limits to fishermen. *Baggots* and *rawners* are terms for salmon that shed their spawn late, or not at all, and then remain in rivers until spring.

Fishing for Atlantic salmon, being so economically valuable, is highly regulated without exception throughout their range. In the United Kingdom many rivers have a river keeper—someone whose main job is to protect the salmon run from poachers. Where truly healthy runs still exist, rich sportsmen may pay a fortune to fish for them, as much as $10,000 per week to visit lodges and to fish in salmon havens like Norway and Russia.

November
American Eel *(Anguilla rostrata)*

Eels do not receive enough respect despite being among America's most amazing fishes. Being catadromous, in November mature eels are *leaving,* not entering, rivers. When an angler lands a writhing three-foot-long eel deep inland, say, in a tributary to Lakes Michigan or Huron, do they ponder the fish's unlikely history—how it got there? If they did, they'd recall that this fish was spawned somewhere in a becalmed area of the Atlantic known as the Sargasso Sea, a more than 2-million-square-mile area

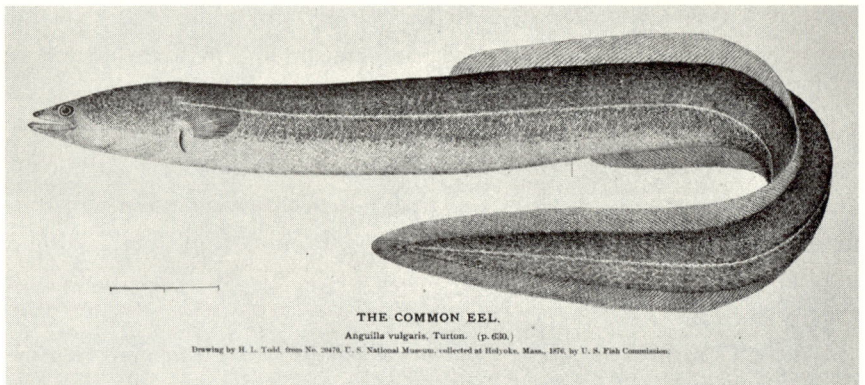

THE COMMON EEL.
Anguilla vulgaris, Turton. (p. 630.)
Drawing by H. L. Todd, from No. 20470, U. S. National Museum, collected at Holyoke, Mass., 1876, by U. S. Fish Commission.

American (Common) eel
NOAA/Department of Commerce

between Bermuda and the Bahamas, encompassing the famed Bermuda Triangle. Other eels on other continents have similar life histories, migratory cycles so simultaneously Herculean and cryptic that James Prosek subtitled his book, *Eels, ". . . the World's Most Mysterious Fish."* He will find little argument.

How do we know this? The American eel has a close relative, the European eel, which spawns slightly eastward of the American eel's zone. Any cognizance of this surprising geographic origin and the fish's subsequent life history before these fish reach Europe as transparent larval "glass eels" was so beyond the imaginations of early peoples, for whom the species were of great importance, that they concocted fanciful tales to account for them. Aristotle was convinced they rose spontaneously from mud, whereas Roman scholar Pliny the Elder believed young eels came from bits of skin that older eels had rubbed off on rocks. Italian anglers believed they mated with water snakes. The seminal angling author Izaak Walton thought they were bred from a particular dew that falls in May and June. The most common theory was that eels arose spontaneously from horse hairs that fell into the water. Yet the truth also is odd.

Glass eels look so different from older eels that in 1856, this life stage was described as a distinct species, *Leptocephalus brevirostris*. A small scientific step was made in 1894 when two Italian professors working at the Straits of Messina in southern Italy made the link between glass eels and elvers, having caught eels of about 70 millimeters long that were transforming to elvers. They concluded vaguely that eels were spawned in the "abysses of the sea," but the location was not specified. It took a Danish oceanographer, Johannes Schmidt, to reconstruct the early life history of eels—and he did it backwards.

What was known was that in autumn, large, mature eels moved down from European rivers and lakes to the sea. These individuals did not return, but in the spring myriads of glass eels, or the next stage, dark little "elvers," appeared on the coasts, eager to reach freshwater. Schmidt asked rhetorically in his 1923 paper, "The Breeding Places of the Eel": "But whither they have wandered, and whence have the elvers come? And what are the still younger stages like, which precede the elver stage in the development of the eel? It is such problems as these that constitute the 'Eel Question.'" This classic paper reads like a detective case as Schmidt systematically analyzes new clues from hard-won data accrued between 1903 and 1921.

In 1904 Schmidt landed a job aboard the *Thor,* a Danish research vessel, studying the breeding habits of food fishes such as cod and herring. One day that spring, a larva of the European eel was captured in one of the expedition's trawls west of the Faroe Islands, between Great Britain and Iceland. Here was a rare clue: Maybe eels spawned far out in the Atlantic? Fortunately for the now-inspired Schmidt, a year earlier he had married the heiress of the Carlsberg Brewery, a Danish company that donated generously to marine research. Intrigued, he began to systematically work toward younger and younger stages. Dragging nets behind schooners, he amassed data showing that the farther from the European coast, the smaller the eels. As he narrowed his search, he also wisely wanted to know where eels were *not found* in the vast Atlantic, and recruited commercial vessels to pull a net for a half-hour or so to give him any eel larvae captured.

In 1913 Schmidt obtained the use of a little schooner named *Margarethe* and sailed it from the Faroes to the Azores to the Newfoundland Banks, and then to the West Indies. But the ship ran aground and was wrecked, leaving Schmidt on St. Thomas with the salvaged eel collections but without a ship. Ever driven and resourceful, Schmidt pressed his fleet of volunteer ships to collect at key locations. When he began in 1903, and for the first few years, Schmidt was capturing eels of about 75 millimeters long; during 1911–12 he went as low as 34 millimeters, and by 1914 he had specimens as small as 9 millimeters. Plotting the data from hundreds of glass eels caught each year, Schmidt found that the farther west, the smaller but more numerous the eels, suggesting he was getting nearer to the mystery spawning grounds. In his paper he asserted that eels must spawn in the southwestern part of the North Atlantic. The document also included a hard-won chart with successively smaller ellipses drawn to show for each species the range of larvae of successively shorter lengths, noting that the smallest must contain the long-sought-after breeding areas for each species. This region is broadly known as the Sargasso Sea.

"No other instance is known among fishes of a species requiring a quarter of the circumference of the globe to complete its life history," he wrote in 1923. "Larval migrations of such extent and duration . . . are altogether unique in the animal kingdom." Subsequent work has shown that the clear, willow-shaped larvae of the American eel drift northward in the Gulf Stream, along the continental shelf, for 170 days or more, depending on growth, currents, and placement along the coast, at which time they metamorphose into glass eels and migrate into estuaries and fresh waters from November through June.

Remarkably, the European eel spawns in the Sargasso Sea, beginning its journey with the Gulf Stream to the Old World from a nearby but separate breeding location. The European eel is nearly identical to the American eel, but develops from seven to sixteen more vertebrae along its much lengthier three-hundred-day larval drift. This simple species delineation recently raised questions about the identity of the eels found in Iceland, roughly halfway between North America and Europe. They had been assumed to be European eels, but there was an elevated frequency of specimens with vertebral counts lower than for typical European eels. Were they genetically pure American expatriates, genetically pure European eels that developed abnormally, or hybrids between the two species? Modern molecular genetics provided the surprising answer: The individuals with lower vertebral counts were hybrids, despite the notion that the two species had separate breeding grounds.

Eels cut a unique figure. They are small-faced, with beady eyes, modest mouths, and a tapering snout. The eel's snake-like body form and undulating swimming style maximizes the species' maneuverability but lowers swimming speeds. Indeed, the eel's maneuverability is legendary. They are virtually impossible to hold in one's hands. Even the "eel grip"—middle finger around and index and ring fingers underneath its body only slightly delays the inevitable escape. They also are living cylinders of nutrition. Whereas most freshwater fish such as pike deliver 300 to 500 calories per pound, even the skinniest eels offer about 700, and they range above 1,600.

Eels live in virtually all waters that can be reached from the Atlantic Ocean that don't have migration barriers—and did they ever migrate. It's been estimated that eels once made up one-fourth of all the fish biomass in rivers of the Eastern Seaboard. Their potential freshwater habitat is remarkably diverse. Eels are found in small headwater streams; large, turbid rivers; black-water swamps; springs and caves, as well as ponds; and from the shallowest to the deepest lakes. Fresh-, brackish, and saltwater marshes are viable habitat as well. It has been recorded that eels could be found under stones in damp meadows near spring-fed rivulets or in metropolitan water pipes. An acquaintance once came across someone fishing for them in a Manhattan sewer line.

One author in the late 1800s, Fred Mather, was struck by how common eels were in the high waters of the Adirondack Mountains, much like in Switzerland, where eels hatched deep in the Sargasso Sea spend their adult lives in waters at elevations as high as 3,000 feet above sea level. Mather not only described eels as occurring in upper headwaters lakes, but also ascending the sixty-foot-high waterfalls separating the Upper Bisby Lake from the remaining Bisby Lake chain. But some waterfalls, such as Niagara Falls, just have too much water falling from too far on high. During a tour of upstate New York, DeWitt Clinton documented that eels were occasionally found as high as fifty feet up its face, attempting to go farther upstream.

The mostly mysterious eel even attracted the curiosity of Sigmund Freud; his 1877 paper, "Observations on the Form and the Fine Structure of the Looped Organs of the Eel, Organs Considered as Testes," was his first published work. Was this a manifestation of a subconscious psychosexual fascination for Freud? Maybe; but sometimes an eel is just an eel.

RIVER-ESTUARY MIGRANTS

There is another school of fishes that make less dramatic migrations between fresh and salt waters. **Shortnose sturgeon** *(Acipenser brevirostrum)* are the little siblings of Atlantic sturgeon. Shortnose sturgeon share the same body shape and physiology, eat similar prey, and also take a long time to mature, with females skipping years between spawnings. But they don't go to sea regularly, as do Atlantics. Shortnose sturgeon are river creatures, largely completing their life cycles in fresh and brackish waters, and only rarely being found outside river mouths. When a shortnose sturgeon is caught in marine waters, it is almost always near shore, likely as it moves between rivers.

A fifty-pound shortnose sturgeon is huge, and most are substantially smaller, likely a consequence of their riverine existences. In *The Song of the Dodo,* his treatise on ecology and evolution on islands, David Quammen repeatedly makes the point that many geographic features are "islands"; islands are not just pieces of land surrounded by water. Rivers are islands in that species that don't venture to sea are restricted to their own rivers. Shortnose sturgeon do occasionally venture between rivers, and so their islands are a little "leaky," but they do not spend substantial time feeding in the ocean, where prey is nearly unlimited. A shortnose sturgeon population must extract its food from the constrained resources of its home river, much as for a marine island population. To put it another way, if the far-larger Atlantic sturgeon were constrained to rivers only, they could exist only at low numbers—and so would be far more likely to become extinct.

Shortnose sturgeon were among the first animals and plants to be listed under the original federal endangered species list in 1967, but not because they were targeted in fisheries. Shortnose sturgeon had been harvested in some rivers—not so much because they were actively sought, but because they were caught along with Atlantic sturgeon and kept as "just sturgeon." In some rivers shortnose sturgeon had access to their spawning grounds cut off by the construction of dams. Data on the numbers of shortnose sturgeon across its range was scarce in 1967, but there was a strong impression among the few biologists who had had contact with the species that most of approximately a dozen and a half populations had plummeted.

These populations were stretched almost as broadly latitudinally as Atlantic sturgeon, stretching from the St. Johns in Florida to the Saint John in New Brunswick. Across that range populations vary by three orders of magnitude in abundance, from the low hundreds in the Merrimack River to the tens of thousands in the Hudson. This raises the difficult conundrum of when to delist it from the endangered species list; is the species still truly in danger of becoming extinct?

As well as the American shad is known is as poorly as science understands its lookalike relative, **hickory shad** *(Alosa mediocris)*. Though called "shad," hickories are more closely related to alewives. Hickories have more of a sweeping belly than American shad, a row of dots running from its gill plate almost halfway down its flank, and fine teeth, unlike American shad. The teeth are a giveaway on its feeding habits; hickory shad is far more piscivorous, living near the coast where it chases small schooling baitfish and grass shrimp. Hickory shad populations are also more southern than American shad, with spawning occurring in rivers from Chesapeake Bay to Florida. At sea, they range to the Bay of Fundy and appear to be moving northward with warming. Unlike American shad they are not highly regarded as food, nor do they grow very large; even three-pounders are uncommon.

If you know **brook trout** *(Salvelinus fontinalis)*, it's likely from upland streams. A coldwater denizen, these char, primitive members of Salmonidae—the family that includes trout and salmon—have been pushed out of many lowland systems by competition with nonnative trouts, and by global warming. Wild brookies persist in these higher-elevation headwaters as small but spectacularly colored vestiges of our heritage; downstream, they more often exist for only days or weeks when stocked in spring before the waters warm for put-and-take angling.

But all the salmonids are ecologically plastic, meaning that in the dynamic glacial and postglacial conditions that characterize their native ranges in the northern hemisphere, they can adapt rapidly to take advantage of favorable conditions. In a goodly share of their range, from Long Island northward, some brook trout went to sea. Not out into the open Atlantic like their salmon cousins, but to estuaries and coastal bays where there was orders of magnitude more food available than in streams—chunky bait-fish like menhaden and anchovies instead of tiny stream insects. Going to sea is optional, though, for coastal brook trout populations; one sibling may remain upstream its entire life, while another may venture to salt water. These trout, known as "salters," lose their vibrant colors to match the duller seawater palette, but they become much larger in the process. Also, a sea-run trout population, deriving its food from broader and richer waters, could be larger than one that draws all its nutrition from a river.

Naturally, such large and numerous trout attracted angling attention. Today Long Island's primordial fresh waters have been diminished by land-fills and a lowered water table. Famous, though, were the salter runs in the island's many swampy drainages in the 1800s, with wealthy sports traveling from as far away as Washington, D.C., Philadelphia, and New York City to exclusive clubs on stream banks to enjoy them. Though there are many versions of the story, Daniel Webster is believed in 1823 to have landed a giant salter of fourteen and a half pounds in Long Island's Carmans River. This catch of a fish that had haunted him for several years and was spied that morning by a slave assigned to watch for the fish occurred either before or while he ran out of services at a nearby church, and was either taken back by train to Manhattan that evening and enjoyed at Delmonico's restaurant, or kept alive in a pen for two weeks for display.

Few fish of any kind were released to be caught again in the sporting credo of the nineteenth century. Dams were also a major factor in the loss of salter populations, as were conversion of some of the headwaters (especially in Cape Cod) into commercial cranberry bogs, with the little streams turned

into drainage ditches. In one case a salter watershed, the Monument River, became the pathway for the giant Cape Cod Canal. Still, restoration efforts have them holding on, on the Cape and farther north, though far from their natural abundances.

The Latin name for the North Atlantic codfish that have nourished Europeans and Americans for centuries is *Gadus morhua*. As important as this species has been to the development of the western world, that's how broadly immaterial its miniature relative, the **Atlantic tomcod** *(Microgadus tomcod)*, has been. A tomcod looks just like an oceanic cod but for some minor details, such as having a proportionally smaller eye. But whereas a codfish can reach one hundred pounds, a one-pound "tommycod" is a monster. Tomcod is a coldwater riverine and bay species rarely found in the open sea. It spawns in brackish water and moves between fresher and saltier waters in an estuary as it seeks its preferred cool temperatures. Sweet-fleshed and a willing biter, it is the consummate dock-fishing species, where it is abundant, but its range has been retreating northward from the Mid-Atlantic States.

At higher latitudes, though, the tomcod is taken quite seriously, at least in one location. In Saint Anne de la Perade, on a tributary to the St. Lawrence River, there is the Festival de la Pêche aux Petits Poissons des Chenaux—an annual two-month-long tomcod ice-fishing festival that attracts 100,000 participants. Anglers fish through holes in the floors of more than five hundred colorful shanties that form an impromptu and even electrified village on the surface of the river. Some hold twenty people and are equipped with heaters, stoves, refrigerators, and bars. Legend has it that the presence of the fish was not known until 1938, when a local who was sawing blocks of ice for his icebox noticed fish frolicking on the bottom of the river. The few ice shanties that were soon built attracted a following, but because there were no open roads, visitors arrived by Canadian Pacific Railroad and even horse-drawn sleighs. Today Saint Anne de la Perade bills itself as the "Tommycod Capital of the World," a designation for which there is no apparent competition.

The **Atlantic whitefish** *(Coregonus huntsmani)* is a little-known species—a glacial relict, i.e., an accident of zoogeography clinging to a sliver

of Nova Scotia and to its very existence. Most whitefishes are inland species of North America and Europe, often found in deep northern lakes. But the group, closely related to the trouts, is plastic enough to have evolved some forms that are salt-tolerant and anadromous. These, however, have not fared well, with a European species going extinct in marine waters. Science knows the Atlantic whitefish from only two rivers, and it was extirpated from one of them, the Tusket, in the mid-1980s. Farmers pitchforking them from a fish ladder there for fertilizer didn't help this population. Neither did acid rain in the 1960s, precipitated into already naturally low-pH waters, nor the introduction of predatory smallmouth bass and chain pickerel. In 1986 the World Conservation Union declared it endangered, only sixty-three years after it was recognized as a species.

The Atlantic whitefish hangs on grimly in the Petite River, just north of the larger Tusket, but for the time being, not as a functional anadromous species. Instead, somewhere between one and two thousand individuals persist in a series of three little water-supply reservoirs, where most of them complete their life cycles. With limited access to the sea, they become stunted; marine migrants once reached fifteen inches, the freshwater landlocks, little more than half that.

I wanted to see this remarkably rare fish in its habitat rather than in a museum jar, and so in June 2007 I met a doctoral student at Dalhousie, Adam Cook, outside a guesthouse in Halifax, and he graciously led us on an Atlantic whitefish tour. First stop was the laboratory where Cook maintained a captive collection for his research. Inside a large tank, bright, torpedo-shaped whitefish flashed back and forth, and I felt privileged to view such a rarity. Next we motored up the Petite River Valley and followed the modest, lovely stream past farms and old mills to the water-supply reservoirs. How had the only anadromous whitefish on the Eastern Seaboard ended up in this fragment of the Canadian Maritimes? The zoogeographic history of life is difficult to prove unambiguously, but the leading theory for the Atlantic whitefish is that just as for Acadian smelt, it survived the Pleistocene on the emergent Georges, Sable Island, and Grand Banks. Robert Schmidt, a zoogeographer of northeastern fishes at Simon's Rock of Bard College, noticed that four New England freshwater fish species had

current distributions best explained by having persisted on the Northeastern Banks Refugium, and then having redistributed to the mainland as sea levels rose. But the Atlantic whitefish inexplicably successfully colonized only the Tusket and Petite Rivers, despite their marine migrations assumedly allowing them access to other nearby rivers.

Cook and I parked at a short stretch of flowing water between the second and third reservoirs. A sign was posted warning against fishing for the whitefish. I asked Cook if he'd ever fished for them anyway. Because it happens to be fun doesn't delegitimize angling as a means of capturing specimens, yet many scientists feel like getting paid to fish is just too shameless to admit. Cook confided he had fly-fished for them, telling me it "was important not to smile" while catching one of the rarest fish in the world. It was difficult to see far into the tannin-stained water, but I could see small fish coming to the surface to pluck insects off the surface—young Atlantic whitefish, the last small parcels of hope for the survival of the species.

Chapter 4

On the Nature of Rivers

Water is the driving force of all Nature.
—Leonardo da Vinci

There are two fundamental ways to perceive, to study, or to simply enjoy a river. If you sit on a rock and watch the flow or stand in a stream with a fly rod and observe drowning mayflies pass you on the water's surface while you watch for trout to feed on them, you are a *Eulerian* observer. But if you hop into an inner tube or canoe and drift with the current, you are in the *Lagrangian* camp. Fortunately, there is no need to take sides; each has its pros and cons, and any ardent river scientist or aficionado practices both.

This simple description glosses over important findings in the 1750s by the Swiss mathematician Leonhard Euler and his Italian counterpart, Joseph-Louis Lagrange. The resultant Euler-Lagrange calculation is a rather imposing set of differential equations that is said to be analogous to Fermat's theorem. But they also developed independent equations to describe flow in these two frames of reference, and both have spawned large bodies of sophisticated work—sometimes combining the two approaches—that allows us to comprehend and to predict how water moves.

Before Euler and Lagrange there was Leonardo da Vinci. The Renaissance master was captivated by water, especially in flowing forms. Da Vinci devoted enormous effort to understanding the most basic properties of flowing water, such as bubbles and vortices. He worked both in his laboratory and in nature, where he studied stream hydraulics using a weighted rod held afloat by an inflated animal bladder. In fact, da Vinci wrote more about water than any other subject. It is our loss, though, that da Vinci never followed through on an outline for a treatise on water found in the margin of one of his papers. Anyone who loves rivers would want to learn about

his thoughts and view his sketches on chapters titled "Of Water in Itself," "Of Rivers," "Of the Surface of Water," "Of Things Moving in It," and "Of Things Worn Away by Water," among ten other chapters. Nonetheless, he left behind many remarkably original insights.

Da Vinci's sketch of a free jet of water issuing from a square hole captures the leonine liquidity but also the sheer complexity of its flow. He likened the motion of the surface of the water to hair, noting two motions: one caused by the weight of the hair, and the other, by the direction of the curls. Or, to put it another way, water has eddying motions, one due to the principal current and the other to the random and reverse motion. Indeed, some hydrologists believe his realization anticipated the well-known Reynolds's formula for the decomposition of turbulence by almost four hundred years. Da Vinci also accurately sketched the pair of the nearly stationary counter-rotating vortices in the wake of an object, commenting on how water wends its way past obstacles, and how large and small eddies are related—observations that presaged important modern hydrological concepts. If these notions appear relevant only to a hydrologist, consider that this is what a kayaker must navigate, and that it is a fish's world, too.

Da Vinci may have been the first to recognize the relationship between earth forms and waterborne erosion generated by these motions, writing in his *Codex Atlanticus:* "Water gnaws at mountains and fills valleys. If it could, it would reduce the earth to a perfect sphere." This physical and progressive wearing and transport of the very vessels of rivers is another link to their biology. Water carves the Earth and, in the process, gives the river its form.

"Study of Water Passing
Obstacles and Falling"
Leonardo da Vinci, c. 1508–1509

Rivers also carry their bounty of minuscule particles—organic and inorganic shavings and flotsam—along with them, resulting in anywhere from chemically uninhabitable to paradisiacally rich and biodiverse flowages.

Some four centuries after da Vinci, there is a substantial but still-emerging science on the nature of rivers. Some river fundamentals: Water flows downhill. Rain and snow falls on the land, and rain and snowmelt run into brooks and streams or percolate underground to emerge as springs. Unless withdrawn for human needs or by intense evaporation in sere landscapes, the waters that run downstream through their catchments are cumulative, as minor tributaries add water to the main stem and as larger trunks merge.

There are exceptions, but the pattern is for the steeper upland slopes at the heads of watersheds to have many small brooks, and for the number of links to lessen downstream as the watercourses become larger. The most upstream rivulets may be ephemeral, visible only during periods of precipitation, and ending at divides—boundaries on the spines of hills and mountains that demark adjoining watersheds. But gravity and water flow make high-relief regions geologically and hydrologically dynamic, and "stream captures" can occur, where an erosively upcutting stream slices into the bed of another, commandeering its flow. Not only does the capturing stream gain more water, but it may acquire new species. This is one mechanism that allows fish and other aquatic creatures to cross mountains and jump drainage basins.

Larger rivers resolutely are "rivers," but smaller watercourses sport a variety of regional names: A "brook" in New England; a "run" in Pennsylvania; a "kill" in New York; a "branch" in the Southeast; and a "creek" out west. The divisions between these terms—streams and rivers—are subjective. But because smaller watersheds normally flow into and contribute to larger systems, catchments by their nature are arranged hierarchically. This hierarchy offers opportunity for a descriptive framework. The most well-known is "stream order," using Strahler's system. In 1952 Arthur Newell Strahler, a geoscience professor at Columbia University, defined a first-order stream as having no tributaries, a second-order stream as formed by the meeting of two first-order tributaries, a third-order stream as formed by the meeting of two second-order tributaries, and so on, a useful but somewhat "dry" way of describing the great melding of waters in which little brooks become mighty rivers.

Describing the physical geography of rivers is far simpler than characterizing the myriad commonalities and differences in ecology among them. Beginning in the 1970s, as the still-young field of ecology matured, conceptual models of rivers began to be developed. The *River Continuum Concept* proposed by Robin Vannote and colleagues has been influential. They noted that the metabolisms of smaller, headwater streams of the first to third orders are dominated by what falls or is carried into the water (like mayflies mating and dying above a stream), with photosynthesis playing only a minor role because of the shading by the tree canopy. But the importance of production from rooted vegetation and plankton increases moving downstream to higher-order links. This in-river productivity becomes more significant farther downstream at even higher orders, but can be decreased by the sunlight-blocking turbidity that often characterizes the lowest reaches of rivers and, especially, estuaries—those important reaches where fresh and salt waters meet. And so many anadromous fishes have evolved to capitalize on the river continuum, depositing their eggs in food-poor waters that also can support relatively few predators, but leaving them in position, after they absorb the nourishing yolk sacs they are born with, to drift downstream into food-rich estuaries.

How are these minerals and other essential chemicals processed within flowing water? The conservationist Aldo Leopold recognized the essential role of the retentiveness of nutrients by rivers when he wrote: "All land represents a downhill flow of nutrients from the hills to the sea." And that this flow has a "rolling motion," meaning that plants and animals "suck nutrients out of the soil and air and pump them upward through food chains; the gravity of death spills them back." That is, without nutrients "spiraling" through temporary captivity in food webs of animals and plants in rivers, these building blocks of life would be carried rapidly downstream and then be shot out to sea.

But today most rivers do not follow the idealized gradients that shape the River Continuum Concept. Hence, the more realistic *Serial Discontinuity Concept,* a corollary which better describes the ubiquitous, less-pristine rivers that are broken up by dams and impoundments. In these kinds of systems, regulating structures such as dams "reset" the river continuum, and not always in the low-order to high-order direction. Because of this, a given stream reach may "behave" ecologically in ways that the River Continuum Concept would predict *should* occur for a different stream order, generating rivers that no longer make ecological "sense."

Add to these concepts the critical notion of scale in ecology. Christopher Frissell and his colleagues at Oregon State University developed a framework of the different evolutionary events and developmental processes that occur at various spatial scales in watersheds. An anadromous fish moving upriver in spring may have only the fierce instinctual drive to reproduce on its mind, but it will "sample" the river as it proceeds at a suite of scales ranging from the river system itself, on the order of thousands of linear yards, created by tectonic forces, governed by erosional planation of the landscape, and persisting for millions to tens of millions of years, all the way down to microhabitat patches of river of less than a yard, created by annual sedimentation, governed by weather-controlled velocity changes, and persisting for weeks to months.

These dynamic smaller-scale changes form much of the basis for the ancient Greek philosopher Heraclitus's famous observation that "you can't step into the same river twice." Recently, some have taken this notion further, saying that even the best Eulerian observer can't step into the same river once! Regardless, watersheds evolve at a series of spatial scales, but not necessarily (and perhaps not even normally) at a steady pace. Yes, over eons erosion appears to be a constant grind, but sediments don't readily dislodge at low flows; it is easier to transport and deposit particles than to first displace them. Interestingly, medium-size particles are most easily eroded: Large ones are heavy, whereas tiny, clay-like particles are "sticky" because of molecular bonding among them. However, the force of water increases geometrically with velocity, meaning that rare but extreme events often have far greater consequences to a river's form than the ongoing but soft drumbeat of average flows.

This "punctuated equilibrium" for rivers, to borrow from evolutionary biology, was well illustrated in an East Coast watershed in 1972. That June, an unusually early hurricane, Agnes, visited the Chesapeake Bay watershed. Though only a Category 1 cyclone in wind speed, it dropped torrential rains of six to twelve inches over a short time, resulting in catastrophic flooding. I still recall driving on a bridge over the Susquehanna River in Harrisburg after the waters receded and looking down at an island to see an aluminum canoe wrapped like a U around a tree some twenty feet up in the air. So much freshwater was flushed into Chesapeake Bay that the seafood industry was damaged for several years. The storm caused the Susquehanna River alone to carry over 31 million metric tons of sediment into the Bay—some thirty times the annual average!

Although one might think the relationship between flow and sediments and, thus, the very nature of rivers is eternal, how rivers functioned was different in the Cambrian Period, half a billion years ago. For decades scientists who thought deeply about rivers entertained a surprising but difficult-to-prove hypothesis: that land plants created the shape of modern rivers hundreds of millions of years ago. Recently, researchers at Dalhousie University strengthened the case for this. The Cambrian's geologic record shows that rivers were shallow but wide, like floods that allowed rainwater to run sheetlike off the barren land. In fact, sediment sizes and distributions suggested that rivers then were one thousand times or more as wide as they were deep.

When these researchers looked at river sediment deposits from the Silurian-Devonian boundary, some 420 million years ago, the patterns changed. The unconsolidated sediments characteristic of the Cambrian appear less frequently, while the depositional footprints of more complex and diverse rivers are seen. There also is more mud, probably due to the enhanced chemical weathering that plants assist. But, most significantly, the shapes of rivers change to highly sinuous, single-thread channels. How could this happen? Plants bind the soil of riverbanks, creating new dynamics between flow and erosion. This was demonstrated experimentally in a laboratory at the University of Minnesota. Alfalfa sprouts were allowed to germinate on the banks of a channel that flowed between multiple sandbars. Over time the system was transformed into one that self-organized into a single-thread channel. The strength of the alfalfa roots was enough to completely change the pattern. Another river researcher commented that these findings "may be considered significant progress in the comprehension of one of the most critical phases in the coupling between physical and biological processes on Earth."

Science proceeds according to well-supported but imperfect paradigms that occasionally are overturned through new findings or new ways of thinking, or a mix of both, as so eloquently outlined by Thomas Kuhn in his 1962 classic, *The Structure of Scientific Revolutions.* A paradigm shift in our comprehension of the form of Piedmont rivers occurred with the publication of a paper in *Science* in 2008 by Robert Walter and Dorothy Merritts of Franklin & Marshall College in Lancaster, Pennsylvania. Until then, river restoration was based on a notion of a characteristic pristine form where water flowed in a single meandering channel through a floodplain. In other words, an archetypical normal river looked much like many assume a healthy river looks today—one main channel with picturesque bends and a sandy or muddy bottom. This form, of course, had become the goal for river restoration.

Examining many lines of evidence, Walter and Merritts showed how wrong that thinking was—how centuries of milldam construction, together with the geophysical cycles they wrought, had radically altered the nature of many East Coast rivers. Walter and Merritts mounted one of those multipronged investigations that are becoming the *sine qua non* of environmental history these days, surveying archived early accounts and maps of milldams along with historical geochemical and geophysical records of river valleys during the period of early land clearing, making their own field observations. Ironically, much of this work was conducted on the same streams and reaches examined in the studies that pioneered earlier fundamental ideas about how rivers behaved through time.

A little milldam history: Europeans had used milldams since as early as 1100 BC, and they quickly applied their know-how in the New World, beginning in the late 1600s. Dams and races that delivered water from the newly formed ponds powered iron forges, furnaces, and mining operations, but most often mills. Indeed, before the advent of steam engines, every mill required a reliable source of dammed water to power it. This resulted in a proliferation of milldams, with peak construction occurring between 1780 and 1860. Walter and Merritts's analysis of 872 counties in the eastern United States revealed more than 65,000 water-powered mills by 1840. Water-powered milling was especially intensive in the Mid-Atlantic Piedmont region, along and west of the fall line. In fact, by the late 1700s the Brandywine Valley had the most notable concentration of milldams in the colonies, with sixty paper mills alone.

This density was achieved despite a less-than-steep gradient—the faster water runs downhill, the more milldams are possible. Even with this modest slope, the investigators found there was one milldam every 1.5 to 3 miles along the Brandywine and its neighboring watersheds. With most milldams ranging between about eight and twelve feet in height, calculations showed that flows would be reduced by 60 percent from about a half-mile to two miles upstream, allowing heavy siltation from the logged and farmed surroundings.

Once the milldams were erected, the accompanying sediments became pale brown and fine-grained, reflecting erosion from the land. These deposits were thickest in the deeper waters near the dams, and thinned upstream from them. Over time, the ponds filled in at the bottoms and sides, with many reaching full sediment storage capacity by about 1850. The investigators repeatedly observed groves of large trees that provided a time marker of up to about 150 years old on valley fill deposits. From then on, the ponds gradually diminished in size and became stable swamps and meadows until the dams breached, causing the waters to cut into the deposits, creating the kinds of simple linear and steeply sided riparian environments we took for normal until this research occurred. Much of today's problematic suspended sediment and nutrient loads in East Coast rivers may be due to this legacy.

Peering below the sediments of the Colonial Era for a view of undisturbed rivers, Walter and Merritts found their natural bottoms contained seeds, nuts, branches, roots, peat, and even tree trunks. These rivers passed through forested wetlands with small branching flows around low vegetated islands that united and separated to form broad necklaces of water. They also contained vastly more woody debris, with natural "snags" and logjams of limbs and branches likely causing new side channels to form, contributing to the dominant braided pattern of flow. The ubiquity of this alternative, natural form was also demonstrated by old maps of European rivers, and today, in the River Lee, flowing through a rare patch of ancient forest near Cork, Ireland. The traditional stream archetype was dead wrong; in the United States two generations of milldam construction inundated and buried presettlement wetlands and drastically altered stream functions and ecology.

But before European colonists modified Atlantic rivers, there were beavers. They also built dams and had been building them for millennia. Still, beavers are often viewed as cute curiosities instead of the remarkable ecosystem shapers they are. Beavers are among the world's most unlikely creatures— oversize rodents imbued with idiot-savant-level abilities to perform hydrological engineering. When I was in graduate school, a fellow student from Taiwan, Moses Chang, refused to acknowledge that the existence of beavers was anything but apocryphal; he insisted that no rodent could perform such dam-building feats. It wasn't until we showed him an actual beaver dam in the Adirondacks, with its carefully woven wall of sticks holding back a substantial pond, that Chang said, "Okay, I admit it; beavers do exist."

In fact, they once existed in extraordinary numbers. Before the arrival of Europeans, some 60 to 400 million beaver were estimated to be gnawing wood from the Arctic tundra to the deserts of northern Mexico. In New England and along much of the Eastern Seaboard, nearly every water body was inhabited and, of course, modified by beaver. But their handsome and useful fur was their downfall, with massive hunting and trapping in the early 1600s sending them into a steep decline, a demand perhaps driven by its coinciding with the coldest portion of the Little Ice Age. Between 1620 and 1630, in Connecticut and Massachusetts alone, more than 10,000 beavers per year were killed for the fur trade. Likewise, between 1630 and 1640 in the Hudson Valley and western New York, approximately 80,000 were killed annually. So great was the taking of beaver, perhaps 50 million in North America alone, that it is hypothesized the resultant drastic reduction from an estimated original 25 million beaver ponds with consequent lowered methane and carbon dioxide discharges instigated a "reverse-greenhouse effect," reinforcing the Little Ice Age and, ironically, creating an even greater need for warm beaver-fur coats.

Today, after a comeback that has brought their numbers to perhaps 6 to 12 million, beavers are often viewed as little more than suburban annoyances whose dams flood backyards. But healthy beaver populations once had a profound influence on otherwise-undisturbed landscapes; their wood-cutting and barrier-building retained sediment and organic matter in river channels, created and maintained wetlands, increased nutrient cycling, and helped to shape associated plant and animal communities. Healthy streams may have fifteen or more beaver dams per mile, each dam holding back thousands of cubic yards of sediment and enlarging the wetted area several hundredfold. One beaver dam, however, became notorious in 2010 when it was seen from space. Most beaver dams are tens to hundreds of feet long, but the beavers in Wood Buffalo National Park in Alberta, Canada, have been working since the 1970s on a structure that now stretches for 2,800 feet. When beaver colonies existed serially along watershed corridors, they were the dominant controlling force across many landscapes. Colonists in eastern North America encountered streams that were broad and ponded, swampy, slow-flowing, and highly productive because of the relentless efforts of forty-pound rodents.

The author of the fly-fishing novel *A River Runs Through It,* Norman Maclean, was "haunted by waters." He is not alone. A river speaks many languages. When I stand in a river, survey the currents, and cast a trout fly to what might be called the edge of a "quasi-stationary counter-rotating vortice," or, more simply, to "nice-looking water," many thoughts pass through my mind, among them, the fact that I am cuing to the end result of many millennia of evolutionary fine-tuning between a remarkably dynamic environment and superbly adapted fishes. A glint of nature worthy of a da Vinci sketch.

Interlude I

A Shad's Journey, circa 1600

The shad wriggles and then pops through the membrane that was her waterborne capsule for ten days as she drifted along as an egg. Her female parent, a medium-size American shad on her second spawning run, had immediately departed the mating area upriver and was working her way back to the sea and its rich food stores to begin another migratory lap along the Atlantic coast. Barely an eighth of an inch long and a feeble swimmer, the shad carries three weeks' worth of provisions—a yolk sac, minuscule but packed with nutrition, to kick-start survival as a speck in the big river.

The shad lives in a soupy world. The flow contains inorganic and organic detritus washed by rains and the melting snowpack from high in the watershed, bacteria that feast on these nutrients, and algae that bloom under the strengthening sun. At her size, though, the world is measured only in inches—flecks of minerals are like boulders; a phytoplankton cell is a beach ball. She drifts with multitudes of her tribe, some hatched earlier than she, some later, and with the eggs and larvae of other fishes—all sharing an evolutionary course where spawning takes place in accord with an instinctual memory of the approximate time and place that often enough in the past led to a good-enough match with the peaking of microbial food production in the river to ensure perpetuation of her kind.

This burst of life near the sunny surface of the river positions the shad in the middle of the food chain. As she finishes depleting her yolk, she begins to graze on minute plankton, obeying the most elemental law of nature—to eat or be eaten. With tens of thousands in her cohort, she coasts along in a defenseless and yet ironical state—there is a sort of safety in numbers—but those high abundances also attract predators. Shad after shad is picked off by perch, sunfish, and minnows as the survivors struggle to eat and grow.

In four weeks the shad has matured considerably, transforming to a diminutive version of her adult form, graduating to larger prey such as insect larvae, and gaining the ability to dodge and outrace some predatory lunges while remaining tightly schooled with thousands of her surviving brethren, including alewife and blueback herring. The school slowly follows the flow and eventually settles into the river's estuary, seeking the densest patches of prey, feeding day and night while attempting to avoid packs of juvenile bluefish and yearling striped bass that slowly decimate their ranks.

As autumn nears and the shad reach finger length, an instinct turns them downriver, to make a run past more and larger hungry mouths, and to make the internal physiological adjustments to full-strength seawater to begin the next phase of their existence. The shad and her comrades pour into the ocean from their natal river along with shad from every midsize and large river on the Atlantic coast, combining into measureless schools that carry on the dual imperatives of feeding and avoiding being eaten. But the predators come, following their own primordial circuits: Codfish that have moved up from the depths to intercept the juveniles as they first enter salt water, great pods of migrating adult striped bass, bluefish, and weakfish fattening before winter, gannets, cormorants—the shad's very existence as a young shad at sea is to swim through a never-ending predatory gauntlet.

Seasons come and go, and the still-immature but rapidly growing shad and her schoolmates move north and south along the Atlantic coast in concert with increasing and decreasing water temperatures. The shad's school absorbs relentless attritions as the field of predators becomes fewer but larger in size, to include sea lamprey, sharks, seals, dolphins, porpoises, and small whales. As the shad migrate, complex biochemical changes occur; many of the females begin to fill their reproductive sleeves with hundreds of thousands of tiny eggs, while also sensing a new pull toward home waters. In this final winter before spawning, the vast schools of shad complete their latest southern loop, but this time the shad is part of one of the many contingents that break off and stream toward their natal rivers as the great shad train decouples.

As the shad approaches the latitude of her river, she detects a faint but familiar odor: the unique mix of scents leaching from the bedrock, soils, and organic matter that characterize the watershed of her birth. This is the signal to tack inshore, where she joins hundreds of thousands of her cohorts from her school and from others, sweeping inland to form a nearly continuous procession of shad homing to the steadily strengthening river scent. Already charged with the calories of countless krill, the shad snatch available food morsels; they must draw on stored energy reserves, and will eat little again until they return to salt water.

The waters shallow and begin to freshen rapidly near the river mouth. The shad's school holds deep in the current, letting the necessary physiological adjustments to the osmotic stress of sea water to sweet water play out. Then they slowly work their way upriver, pausing here and there to conserve energy in quieter river reaches. The shad's abdomen swells as her hundreds of thousands of eggs begin to hydrate in preparation for spawning. For several days she and her school fight the escalating current to gain a few hundred yards to a mile or two, and then pause, the next advance often a response to the sudden departure of the school preceding it. Mingled with them are tens of thousands of alewives and blueback herring, holding in shallower waters as they aim for the river's more-distant headwaters and tributaries. Then the first few shad moving downriver are passed; though it's early in the run, some now gaunt-looking individuals have already spawned and, exhausted, are riding the flow. An otter shoots by and seizes a still un-spawned shad, instinctively knowing it is a far richer meal than a fish that has completed its mission.

Pressing farther upriver, the channel narrows into classic riffle and pool habitat, and the entire spawning reach becomes crowded with restless shad. A portion of the shad's school is squeezed close to a strategically placed rock weir and a whoop is heard as a Native American fisherman yanks on a net and captures several of her cohorts. Cat's-paw winds ruffle the surface of the pool when a large dome of high pressure from the north brings a sharp chill, and the water's temperatures plummet below the innately sensed threshold for spawning. The waves of shad come to a halt and the fish wait, patiently holding in the deepest, most quiescent waters available, both to conserve energy and to avoid the ongoing swoops by ospreys and eagles that key to these fish as they have for seasons immemorial to feed their own young.

The days and nights become warmer as the winds weaken, and on the third evening, they shift to the southwest, bringing even milder air and, not long after, rising river temperatures. The shad begin to stir. Because they produce milt longer than females produce eggs, males arrive first on the spawning grounds. And because it's still early in the run, there are still many more males than females. The shad does not lack for suitors.

Dusk fades to night and the pool becomes alive with milling and swirling shad. Males chase females, bumping them, urging them to release their eggs. A young male pairs with the shad and nudges her continuously until she begins to stream roe. They both quiver as they extrude milt and eggs that cloud the water, mixing with the products of hundreds of other shad, the activity becoming so pitched that it becomes orgiastic, with shad simply releasing their seed into the suddenly opaque waters. The shad rests during daylight and repeats this performance for two more nights.

Wearied and drained of eggs, the shad and other spawned-out individuals group and begin to ride back downriver. Along the way these back runners pass the first schools of striped bass and sturgeon that have entered the river to spawn, the predatory bass eyeing the shad but chasing the smaller river herring. The shad let the flow assist them until they reach the river's lower reaches, where they must swim or hold in the flow against the incoming tide, but can then ride the ebb tide seaward. These estuarine waters contain zooplankton and small shrimp, and the shad begins to feed again to replenish her depleted energy reserves. The salt seems foreign at first, but again her superbly responsive chemistry makes the requisite calibrations. With a sustained burst the shad drives forward into the sea, having completed the grand life-history circuit of an anadromous fish.

Chapter 5

On Natural Abundances: Remembering Not to Forget

We transform the world, but we don't remember it.
—*Daniel Pauly*

That society largely fails to recall what it once had but has since lost is indisputable. This process of forgetting occurs in a pattern so regular that a prominent fisheries scientist, Daniel Pauly, described it and named it the *shifting baselines syndrome,* initiating a powerful new paradigm that has resonated throughout the conservation realm. Wise Thoreau foreshadowed its essence a century and a half earlier. In *A Week on the Concord and Merrimack Rivers,* Thoreau wrote, "Dim visions we still get of miraculous draughts of fishes, and heaps uncountable by the riverside, from tales of our seniors."

Dim visions indeed. Dim visions of unimaginable plenitude are often all we are left with in today's world. And sometimes not even that.

Pauly was raised far from the sea, in Switzerland, having what he called a Dickensian childhood as a live-in servant. After graduate work in Germany he spent the early portion of his career in tropical locations where—being half-black—he thought he might be more comfortable. Although Pauly made important contributions in developing quantitative methodologies in managing fisheries, he viewed the problems of overfishing broadly and examined all available information. Pauly's seminal paper on shifting baselines was a scant one page long, more of an essay than a research article. In "Anecdotes and the Shifting Baseline Syndrome of Fisheries," Pauly wrote, "Essentially, this syndrome has arisen because each generation of fisheries scientists accepts as a baseline the stock size and species composition that occurred at the beginning of their careers, and uses this to evaluate changes. When the next generation starts its career, the stocks have further declined, but it is the stocks at that time that serve as a new baseline. The result obviously is a gradual shift of the baseline, a gradual accommodation of the creeping disappearance of resource species, and inappropriate reference points for evaluating economic losses resulting from overfishing, or for identifying targets for rehabilitation measures."

This consummate number-cruncher had somehow been impressed by tales from days of old, anecdotes such as those provided by a colleague whose grandfather had caught bluefin tuna in the North Sea, a place they are absent from today. Pauly argues that although this information is "as factual as a temperature record," modern fisheries science has no formal means to incorporate it into its investigative framework. Pauly also believes that integration of fisheries history would help us to overcome—in part, at least—the shifting baselines syndrome, and to evaluate the true social and ecological costs of fisheries.

In effect, Pauly describes a ratchet, or a slippery slope where, after a while, you can't see the hilltop; later, not even much of the grade above you. And perhaps you even stop caring. But he also helped to inspire a response, providing an influential prod to the rapidly growing discipline of *historical ecology*.

What is historical ecology? Why is it useful? Historical ecology is the visioning of lost ecologies through one or more methodologies, all of which strive to be rigorous in the face of often-substantial uncertainty. Pursuing historical ecology (like so much of science) can be satisfying simply for the sake of the knowledge gained. But historical ecology has a strongly applied benefit, by using the past to better manage ecosystems for the future.

One of the intrinsic ironies of historical ecology is that the earlier the information source, the more valuable, but also the more tenuous it is likely to be. Nonetheless, part of the appeal of historical ecology is how it draws from so many potential sources. These sources stretch across their own spans of time but also overlap the records of others, sometimes allowing synergies among them. To go back thousands of years, we have paleontology. Picking through the fossil record reveals natural rates of extinction; some of these losses are normal (i.e., there is a "background" extinction rate), but comparison with the present shows extinction rates one thousand times higher. However, paleontology is not just old bones. Paleontologists can take cores of sediments or coral reefs to extract layered records of accumulation that reflect past ecological conditions. Chemical and radiological investigation is possible, too. For an instance involving anadromous fish, researchers used shifts in the abundance of a nitrogen isotope in sediment cores to show large fluctuations in the abundance of Pacific sockeye salmon in conjunction with changes in climate and coastal ecological productivity over 2,200 years—thereby "setting" a baseline for natural, long-term variation.

Surprisingly, genetics also allows for reaches far back in biological history. The modern molecular revolution combined with brute-force, computer-based modeling approaches permits analysis of important questions that could only be speculated on thirty years ago, such as whether the genetic diversity of a whale species today is now radically reduced because of whaling, or how large its populations were prior to hunting. When applied to North Atlantic whales, some of the results have been striking: One study suggested that minke whales are at 56 percent of their original size, but that fin whales are down to 15 percent, and humpback whales only 4 percent.

Archaeology is more recent but takes us back hundreds to thousands of years. Layered heaps of garbage, or "middens," reveal changes in human effort, such as when the species composition in remains left by a coastal people shifts from inshore to offshore prey. They may also provide a rich glimpse of changes in the availability of natural resources, sometimes including diadromous fishes. In the southern Baltic Sea, remains of European sea sturgeon dropped from 70 percent of fish consumed in the seventh to ninth centuries to 10 percent in the twelfth and thirteenth centuries, likely the early part of a trend that continued to the near disappearance of the species from European waters today.

The written record overlaps the archaeological, but for historical eco-logical purposes is limited to hundreds of years. On the plus side, writings come in many forms: reports, logbooks, maps, diaries, letters, and even cookbooks and restaurant menus. Based on records of bankers, finan-ciers, and tax collectors, researchers were able to reconstruct a time series on Mediterranean tuna landings in trap nets between 1650 and 1950 that demonstrated a relationship to temperature changes. However, they found that after 1950, overfishing swamped any climate effects. Even photographs have proven useful; for example, an investigator gleaned insights by look-ing at celebratory snapshots made at a Florida dock of catches by a char-ter sportfishing boat between 1956 and 1985. Although the lengths of the catches were quantified by comparison with the dimensions of objects in the photos, the significance of the chronology was obvious simply by view-ing it. Early photos contained numerous specimens of goliath grouper—brutes weighing more than one hundred pounds each; later, goliath grouper became scarce, but respectable catches of other gamefish were made; and by the end the catches that tourists proudly posed with from their charter trips would have barely qualified as bait in the 1950s.

Fisheries science has matured over the past two centuries, with the nature of its data becoming more sophisticated along the way. Data on fish-eries are gathered in two ways: either from the catches themselves, or from independent surveys of the status of the populations actually fished. The problem is that the start of most fisheries data collection did not occur until long after the actual fisheries began, reducing their value as baselines. But there are exceptions: From comparison with a baseline survey in 1961, the biomass of large predatory fish in the Gulf of Thailand was shown to have fallen by more than 90 percent.

On a shorter time frame—decades—is historical memory. This is one source that makes some more-obsessive scientists nervous, given its pliable nature. Yet potentially there is much unique observation and experience con-tained in human minds that unfortunately will never make it to the written record. A comprehensive reaping of fishermen's memories was conducted in the Gulf of Maine by a former commercial fisherman, educator, and sci-entist, Ted Ames. Ames, who went to sea at age six with his grandfather, became so discouraged with the state of Maine's inshore fisheries, especially for codfish, that in 1990 he sold his forty-five-foot trawler. But he saw an opportunity to perhaps get the government to stock cod on their historic spawning grounds—if those spawning grounds could be identified. Ames interviewed twenty-eight of "some of the most wonderful . . . old codgers you could ever imagine." He then painstakingly plotted information on the locations of the seasonal cod-fishing and spawning grounds, and characteristics of the fish on the many peaks and gullies that form the rugged underwater landscape off the Maine coast. From this he was able to reconstruct the fundamental biology of the many localized and since-extirpated local cod stocks. So impressive was this work that he was awarded a MacArthur Fellowship for it.

An especially sensitive antenna of environmental variation is long-term ecological monitoring—after all, detecting change is its very purpose. But "long-term" is actually "short-term" on the scale of historical ecology; recognition of and commitment to the need for this is not that common, nor does it usually extend further back than decades. The results of annual multimillion-dollar surveys of the Hudson River that have been funded since the 1970s by the power utility companies that use river water have shown dramatic shifts in the ecology of the river and its diadromous fishes over that time, in response to human-induced alterations such as the institution of the Clean Water Act and the accidental introduction of zebra mussels. The level of detail in the data these programs generate is superb, but they only function as references for future change; rarely do they serve as natural, predisturbance baselines. Still, their value increases with each additional interval accomplished; the longer the time span, the greater the potential for separating random fluctuations from meaningful ecological signals.

Having a data source is not always enough to perform useful historical ecology analysis; it also requires a conceptual framework, particularly when different forms of information are combined. Information from a single point in the past is useful, but not ideal—such "then-now" comparisons ignore all that occurred in the interim. Much better is a time series where information is available at additional points between the earliest and latest instances, allowing meaningful comparisons with ongoing factors.

"Hindcasting" uses simple population models to estimate aspects of ecology backwards in time—like future predictions, but flipped over. For instance, models of the "carrying capacity": how many fish could be supported out of twenty-one North Atlantic cod stocks. Results showed that they now ranged between 0.1 and 16 percent of carrying capacity, a truly abysmal set of results. One more conceptual approach to historical ecology is "space-for-time" comparisons, such as the still largely unaltered River Lee in Ireland. This was the method used by Alan Weisman in his provocative book, *The World Without Us,* a look at how global ecologies would change in the sudden absence of humans. Those rare places or "spaces" that exist today with little human influence serve as proxies for the effects of time, a prime example being the 400-square-mile Korean demilitarized zone. Almost six decades with virtually no human presence there provides a glimpse at what the terrestrial ecology of eastern Asia would look like (and what it did look like) without humans. In essence, space was swapped for time.

Historical ecology is just beginning to mine potential data sources. Challenges will include stitching together information from diverse wellsprings to re-create reliable baselines and subsequent trajectories on particular ecosystems. But there is a fundamental issue in doing so, in that even natural ecosystems aren't static; climate changes, invasions by new species, and extinctions cause shifts. This means that choosing "pristine" baseline conditions as a reference are somewhat arbitrary; they will depend on the time point selected. This, however, is more of a philosophical issue than a practical one. For most lost ecologies, that any reconstruction is possible is fortunate; resource managers take what they can get when reaching backwards into the murky recesses of time. Restorations that occur today are usually so constrained by the many anthropogenic changes that have occurred in the interim that there is no potential for exact re-creations of historical ecologies anyway, only a shift to vague approximations.

Nonetheless, the now-lively world of ecological restoration is hungry for historical information on the ecology of the locations targeted. And ecological restoration is also an evolving discipline, and one that is divided into competing philosophies concerning particular goals in relation to knowledge of baseline conditions. All are "re-" words: *"returning"* the original

Boys admiring Atlantic sturgeon catch, Saint
John, New Brunswick, c. 1960s
Provincial Archives of New Brunswick

condition, "*re*storation" of ecological integrity, "*re*wilding," and "*re*naturing." All involve taking a step forward by looking backward. Returning the original condition is literally going back to the baseline conditions, something that usually is practicable only if there has been minimal change from the baseline. Restoration of ecological integrity is the basis for many US government programs. That is, the notion that a return to original conditions is no longer possible or desirable, but that a recovery to a proper ecosystem structure and function is, albeit not the original one.

Rewilding may be the most interesting alternative. The idea behind rewilding is to reestablish the original setting so that natural selection guides the future state of the system. If this sounds like returning to the original condition, well, it sort of is, but with the important distinction that the motivation and logic are different. In rewilding, rather than re-creating the past and all the complex interactions that this entails, the aim is to restart the evolutionary process for that ecology under prior conditions. The fourth alternative, renaturing, is in origin a European approach. To renature a place is to design an ecosystem for nature and for people, with the two being equal partners.

All four restoration approaches are preferable to maintaining a severely compromised status quo. Before some ecological restorations occur, these alternatives are argued at length; in others the goals are obvious, and the managers just plunge ahead. What diadromous fish restoration occurs on the US Eastern Seaboard proceeds largely under the philosophical umbrella of restoring ecological integrity, but many of these efforts have failed to bring back any semblance of the original runs that were fundamental to the ecological integrity of the rivers. These restorations also have proceeded with some general knowledge of historical ecology, but rarely integrated along long-term timelines. That is, for a given river, managers may have a vague understanding of Native American fisheries; they may be aware that colonists harvested certain species; and they may have access to modern fisheries records, but detailed syntheses are rare. And the shifting baselines syndrome tends to cause them to simply hold the fort against further loss. I would argue that in reality, diadromous fish restorations on rivers with large dams, for instance, have been a form of renaturing, but one in which the fish are junior, far lesser partners, to people.

The shifting baselines syndrome, as important a paradigm as it is, is not the complete story in the current poor state of diadromous fish runs, because the syndrome is explicitly about the limited vision of the fisheries scientists. The public also is complicit because of a parallel lack of vision. The slow, "invisible collapse" of these fisheries meant that the role these fishes played in the lives of early river-valley residents and the people they traded with faded. No longer did family members fish the rivers for food and for market; no longer were these fish on the dinner table; no longer did residents hold festivals celebrating their return; no longer did they mat-ter—and so they were forgotten. This forgetting at the public level has been termed "intergenerational amnesia." One consequence is that the loss of memory of these fish, even while still present but declining, resulted in a loss of societal standing for them, so that their absence provides a sort of "permission" to continue adulterating rivers.

We need to remember. Thoreau provided a gentle prompt nearly two centuries ago: "It cannot but affect our philosophy favorably to be reminded of these shoals of migratory fishes, of salmon, shad, alewives, marsh-bankers, and others, which penetrate up the innumerable rivers of our coast in the spring, even to the interior lakes, their scales gleaming in the sun; and again, of the fry, which in still greater numbers wend their way downward to the sea."

Karin Limburg and I proposed a label for the combination of the shifting baselines syndrome of managers and the intergenerational amnesia of the public, calling the cycle through both "ecosocial anomie." The phrase has not caught on, and I now wonder if the world needs another arcane term for these phenomena. The blogger J. B. McKinnon did react to our proposal by developing a chronology of the various definitions that have emerged over the years, including those mentioned above and a number of related ones. But the title of his commentary summed it up best: *Forgetting We Remembered We Forgot*. We need to do better. We need to reshift those baselines.

Chapter 6

Spearfish Moon

Understand that the river knows its destination.
—*Thomas Banyacya Sr., Elder of the Hopi Nation*

American Indians were not exactly smitten with the foreign invaders. They found the colonists oddly dressed, many of the men having peculiar blue eyes that peeped out of bristly, animal-like hair that encased their faces. "Europeans, Indians told other Indians, were physically weak, sexually untrustworthy, atrociously ugly and just plain smelly." In fact, Europeans were dirty, too, unbearably so; many had never taken a bath in their entire lives. The use of handkerchiefs for nasal emissions disgusted the Indians. They also characterized the interlopers as "irritatingly garrulous, prone to fits of chicanery, and often surprisingly incompetent at what seemed to the Indians like basic tasks." And yet, the presence of Europeans offered an irresistible chance to trade for things unobtainable in the New World, items such as copper kettles, steel knives and hatchets, and beads of colored glass.

The beginning relationships between these two peoples—about as different as can be—were tumultuous to say the least, varying from wary tolerance with offerings of assistance by the Indians, to outright war. Schoolchildren in the United States are initiated into this history with a highly mythologized account of the first "Thanksgiving," a large gathering of Indians and Pilgrims that occurred sometime in November 1621. A noteworthy participant was Tisquantum (aka, "Squanto"), who spoke good English because he had been kidnapped from his Patuxet village in 1614 and brought to England for several years before returning and living with the Pilgrims as a counselor, in effect. Among the things Squanto taught them was how to catch eels.

Estimates on numbers of pre-contact Indians are terribly imprecise; across what later became the United States, they've ranged from 1 million to 18 million. Sixteenth-century New England may have been home to 100,000 Native people or more, a figure that was slowly increasing. Native Americans inhabited eastern North America for many thousands of years, but not as far north as New York and New England. Because of glaciation and the bleak landscape it left behind, it is likely that the number of Indians in the Northeast was low until the landscape had passed through ecological succession from tundra to forest, in the process becoming more human-friendly.

Along the Eastern Seaboard, coastal Indians often lived a fifteen-minute walk inland to avoid direct exposure to storms and tides. Communities were not fixed, joining and splitting over time, though larger, more-permanent villages often persisted in major river valleys. Their mix of agriculture and foraging also was fluid, reflecting the particular resources nearby. Foraging included some still-debated level of reliance on fish that ran between rivers and the sea.

Look at an aerial photo of Paterson, New Jersey, and you'll see a representative snippet of why New Jersey is the most densely populated state in America. In fact, Paterson's 150,000 residents are sufficiently packed to render it the nation's second most crowded city after New York. Through the heart of Paterson flows the Passaic River, an urbanized waterway that farther downriver features a Superfund site with high levels of dioxin. The view shows both shores of the Passaic to be lined with industrial buildings, but there is also a curious feature clearly visible in the water that stretches almost one hundred yards from bank to bank—a V-shaped stone weir. Built to harvest migratory fish some unknown number of centuries ago, it persists as incongruous and quiet testimony to a time when the Passaic Valley was verdant and its waters rich with life, including eels, sturgeon, shad, smelt, and striped bass.

It is indisputable that diadromous fishes contributed a portion of the dietary needs of Native Americans along the Atlantic Seaboard. Exactly how, though, is the devil in the details. Diadromous fishes seemingly would form a great and predictable food source, yet the evidence does not indicate the kind of reliance on them shown, for instance, by the salmon-fishing tribes of the Pacific Northwest. Were diadromous fishes only a minor and ancillary resource for East Coast Indians, or is this persistent notion an artifact of the peculiarities of fishing methods, the usage of the fish, and of archaeological preservation? These peoples had no written language and so they left no written accounts. But two historical ecological sources provide a substantial, if imperfect, record of their relationship with these fishes: archaeological remains and the observations of Native Americans by early colonists.

What is known is that Native Americans caught diadromous fishes with great knowledge of their natural histories within rivers, and with ingenious usage of the natural materials they had available to them. The Lenape called the month of March, when shad first appear, *Chwame Gischuch,* which translates to the "month of the shad." And the moon of the peak shad migration in April was known by the Algonquian as the "Spearfish Moon." Fundamental among Indian techniques was the construction of weirs, such as the Paterson weir, either rock or brush linear obstructions that steered migratory and residential fishes into holding areas where they could be gathered. But then, why are the locations of so few of these known today? Due to sea-level rises over the previous 3,000 to 4,000 years, prime downriver locations such as river mouths and in estuaries no doubt were inundated. The wooden stakes of brush weirs must have almost always decayed, except under extraordinary conditions. Many rock weirs would have been broken up by centuries of flows. Thoreau referred to them in *A Week on the Concord and Merrimack Rivers,* both to weirs he knew still existed on an upstream tributary, and when he drifted through an "old battle and hunting ground . . . the ancient dwelling place of a race of hunters and warriors," where "weirs of stone" together with arrowheads and hatchets lay concealed in the mud of the river bottom.

"Their Manner of Fishynge in Virginia"
Theodor De Bry, 1588

Though weirs had not received much attention from archaeologists, Allen Lutins, a master's student at Binghamton University, became engrossed by them. He found that weirs were even more integral to Indian life than had been realized. The Scandinavian explorer Peter Kalm noted tidal weirs targeted at eels in the St. Lawrence River estuary that were made of twisted osiers, "so close that no fish can get through them," with a net at the portal made of twigs or yarn. Weirs were widely used within the Iroquois Confederacy. Dablon, a Jesuit missionary to the Oneida, wrote in 1670 with admiration that "Our savages construct their dams and sluices so well, that they catch at the same time Eels, that descend, and the Salmon, that always ascends." Some of these weirs could hold five or six hundred eels at once.

The best-known prehistoric weir along the Atlantic Coast is the one at Boylston Street, discovered in 1913 by subway workers, that lies buried twenty-nine to forty feet under the city of Boston. There, about 65,000 stakes interwoven with brush occur over two acres. Despite its great size, a recent interpretation is that the stakes constituted numerous small weirs constructed over as many as fifteen centuries. Weirs abounded in Virginia and North Carolina. Descriptions from the 1580s were lacking as to detail, but clues to the form of these "reed" weirs are provided by John White's watercolors from 1585 and 1586. Also, elaborate stone weirs were known from near inland waterfalls in Virginia. Rocks were set across the river, leaving one, two, or more spaces, or "trunnels," for the flow; at their mouths were set ten-foot-long, cone-shaped baskets. Fish would swim into these and become wedged and held by the current.

The Paterson weir, one of eleven noted on the Passaic in 1913, today stands alone on that river. Entropy from the force of endless flows, natural decomposition, and ignorant or unconcerned construction activities all helped to erase the physical records of this important way of life in the Passaic and up and down the Eastern Seaboard.

But weirs were only one way to pull a fish out of water. Archaeologists have found half-pound sinkers made from flat river stones to hold down the bottoms of nets used for shad and other species, in addition to anchor stones weighing as much as eighteen pounds to secure the ends of nets that sometimes were prodigious, as long as 400 to 500 feet. Funnel-shaped fish baskets made of splints were anchored in strategic locations in riffles, with the fisherman spooking and steering shad downriver into them. Sometimes teams would drive fish downriver into a weir-based trap using a line made from vines held crossways in the current and laced with brush; this technique could produce a thousand or more shad and other fish in half a day. The internationally ubiquitous fish hook was made east of the Appalachians from bone or dried bird claws. Fish were speared and shot with bow and arrow. Mild poisons from sources such as walnuts were used when fish were trapped in pools during dry seasons. The Tuscaroras of North Carolina built fires on riverbanks at night to attract fish, where they were harvested with reed baskets or shot with arrows. At night, Indians also would build fires in hearths in their canoes, or simply hold out torches to dazzle and mesmerize the fish before impaling them.

The largest of the anadromous fishes, sturgeon, were captured by Native Americans in a variety of ways, including with weirs and nets and with green-cane spears in Georgia, and harpoons and torch lights at night by the Abenaki in Maine. Some even used a highly direct and physical approach, as reported in 1705 by Robert Beverly: "The Indian way of catching sturgeon, when they came into the narrow part of rivers, was by a Man's clapping a noose over their tail, and keeping fast his hold. Thus a fish finding itself entangled, would flounce, and often pull him under water, and then that man was counted as a Cockarouse, or brave fellow, that would not let go, till with swimming, wading and diving, he had tired the sturgeon, and brought it ashore." Today we would fashion a reality television show from this; back then it was a meal.

The significance of diadromous fishes to Indians is shown by their inclusion in artwork. One long-lasting iconographic form is the petro-glyph—images that are carved into the surface of rock faces. Along the Susquehanna River, which ran with astounding numbers of American shad, there are sites once on river islands that speak to the importance of these fishes. At Bald Friar, a traditional Indian fishing location at a passable falls, every large boulder had glyphs. Twenty-one of fifty-three cataloged figures are human-head-fish-effigies—fish with blended human and animal facial features. Anthropomorphic faces illustrate that the fish are ensouled—given a spiritual status in an animistic understanding of the world. Rather than swimming across the rocks, these shad spirits look directly at human beings, eye to eye. In 1926 some of the glyphs were removed by researchers and curio hunters. The remainder sits on the bottom of a reservoir. Any magic they once might have worked has been defeated by the giant, shad-migration-crippling Conowingo Dam.

With this substantiation for a variety of Native American capture techniques for diadromous fishes, how important were they to their total
diets? There is a tendency when considering Native American harvests
of anadromous fishes to focus on the dramatic runs of adults during the
spring upriver migrations, as the colonists did, processing and hoarding the
catch for later use. Bill Schindler, then a doctoral student in anthropology
at Temple University, challenged this orthodoxy, suggesting it was a view
biased by our contemporary fishing practices, our notions of suitable sport
fish, our ethics, and our fishing regulations. Schindler's revelations sprang
from his having taken a close look at the remains from Indians of the Middle
Woodland Period in the Delaware River Valley, right in the heart of the
latitudinal range of diadromous fishes on the Eastern Seaboard.

Earlier interpretations concluded that Indians needed to amass great
amounts of focused labor from mid-March to mid-May to take advantage
of migrating adult anadromous fishes. But experimental research showed
just how long it would take to process a harvest of even just 1,000 fish
using stone tools and primitive methods. The answer: a *long, long* while.
Two hundred "people hours" were needed to scale and butcher 1,000 fish;
twenty people hours were required to construct a drying rack to handle
that many; and it took 250 hours to collect enough firewood to keep several
small smoke-producing fires burning twenty-four hours a day for ten days,
to deter insects from the drying flesh. This was quite a labor force to sum-
mon for a brief period from a thinly distributed population.

Schindler argued instead that diadromous fishes of all edible life stages
had a prolonged availability that made them a much more regular food
source that did not require such a punctuated effort. About ten diadromous
species occurred in the Delaware River. Of these, only two automatically
died after spawning as part of their life cycles: sea lamprey (in the river)
and American eel (at sea). The others experienced some unknown and partial mortality through the stress of making a spawning migration. Thus,

an upstream run of American shad numbering a million and experiencing 50 percent mortality would still leave available a half-million adults that would filter out of the river for a month or two after spawning. The shad alone would extend their availability far beyond that point, however; later, in summer and in autumn, their progeny, possibly numbering in the millions, would also pass downstream as small edible morsels. Taking the shad together with the life stages of all of the other diadromous species and plotting the timing of their presence in the Delaware River, Schindler showed that diadromous fish were readily available from March into November, often as more than one species or life stage. There was, in fact, no need for a sudden spring labor muster, and processed fish only needed to be stored for the cooler, winter months. Archaeological evidence did indeed indicate fishing sites were used in multiple seasons.

Long before Europeans began working on the Eel Question, and before Squanto educated the Pilgrims in the ways of the eel, many Native American tribes were feasting on them. Excavations of prehistoric archaeological sites suggest eels were the main vertebrate—at least by abundance—exploited by Native Americans living in the Lake Ontario and St. Lawrence River region. That eels were important to Indians was apparent to early colonists; in 1535 Jacques Cartier noted that the Iroquois of Hochelaga (present-day Montreal) had large vessels in which they kept live eels they smoked during summer to survive on during winter. Eels were so significant to Indians of the St. Lawrence region that at least three of the New York Six Nations had an Eel Clan—the Cayuga, Onondaga, and Tuscarora—the only clans in the Iroquois League named after a fish. Indeed, the name *Algonquin* may be related to eels.

To Algonquins, eels were sacred, and no part of one was wasted. The Indians' hair shone with animal fat to protect it from the sun and wind and to ward off insects; both bear and eel fat were employed, but eels were far more available. Eel skin was used decoratively for hair strings. Eel skin tightens as it dries, which also made it useful for bindings. It was used to sew moccasins and clothing, to secure sled frames, for tying spear and harpoon points to poles, and for applying splints to broken bones. The hearts, livers, and heads of eels were sometimes buried as offerings after successful catches. Eels could also be thrown into a fire to appease the Devil. In times of famine, smoked eel skin was the food of desperation. The old story of starving aboriginal people eating their moccasins for the leather is only partly correct. Fresh eel flesh was applied to clothing to waterproof it. Buckskin fringes evolved as a practical design to absorb more eel fat, as well as to discourage insects. Because of this, in times of dire hunger, treated buckskin offered a few precious calories of nutrition.

Indians also used eels in trade. Samuel de Champlain in the 1620s reported the indigenous people to be skilled at catching eels, but also hard bargainers in trade, with his hungry men giving their coats and other possessions for the fish, and Champlain himself buying 1,200 eels at the rate of ten eels for one beaver. While near Onondaga Lake in 1655, a medium-size water body within the Oswego River watershed, missionary Father Chaumont wrote: "[T]he eel is so abundant there in the autumn that some [Indians] take with a harpoon as many as a thousand in a single night." Catches of eel were so high in the late seventeenth century within the St. Lawrence drainage as to be considered "an infinite quantity." Further reports from that period estimated eel were once considered the most common fish, along with salmon. Indeed, the lowly eel may have been more important to New England and St. Lawrence region Native Americans than more-glamorous diadromous species such as salmon, shad, or sturgeon.

One North American anadromous fish species did some surprising precolonial wandering that put it in the hands of European fishermen centuries and even millennia ago. This most extreme case of "exploratory behavior" by an anadromous fish was first hinted at by remains unearthed at the medieval archaeological sites of Ralswiek and Wilhelmshof, along the German Baltic coast. These digs suggested a remarkable story—that a few Atlantic sturgeon from North America had crossed the Atlantic Ocean by the eighth century AD and colonized the Baltic region, a tale suggested by notable differences in the surface sculpting of the bony scutes that armor a sturgeon's flanks and which survive well in deposits. I worked with researchers who used genetic evidence along with the scutes to conclude that the colder conditions of the Little Ice Age allowed the American sturgeon to displace the native European sturgeon, a species that favors warmer waters. Further genetics research extended the amazing narrative; modern analytical techniques indicated that only about ten American individuals founded the new colony, and that it was clear that there had been some hybridization between the sturgeon species of both shores.

Once the radical notion that American sturgeon had colonized Europe had taken hold, the French reexamined their archaeological collections. Lo and behold, the American fish had arrived even earlier, in fact, much earlier, on their coast. Three sites proved to be of interest, all on the Atlantic shores north of Bordeaux. Together, they showed that the American Atlantic sturgeon was present in French rivers and along the coast at least five thousand years ago, and that they persisted—and coexisted with European sturgeon—until at least the second century AD.

These findings energized the European sturgeon community, which had seen expectations shrink for restoring their native species as its numbers dwindled in the one population left in the wild. Today's conservation biology strictures do not allow purposeful introductions of nonnative species. These findings suddenly showed that the American Atlantic sturgeon was native! Already, the American species is being cultured in German facilities and stocked in the wild, with most European sturgeon restoration hopes resting on it.

How much had the diadromous fishes of the Eastern Seaboard declined by the time Europeans appeared in the New World? That is, even though the colonists were sometimes awestruck at the abundances they encountered, how much greater would they have been in the absence of any fishing whatsoever? The question has never been answered, and even a coarse estimate would need to involve many dubious assumptions. By the time Europeans reached North America, fishing may have been in decline by Native Americans, owing to increased dependence on agriculture. It seems likely that the longer period of harvest of various life stages of a large number of species by low-density Indian populations, some of whom were hunter-gatherers who were not major fish eaters, may have left the anadromous fish running silver in nearly full primordial glory when the first European colonists arrived.

What is known, though, is that Native American populations themselves declined quickly after European contact. Squanto returned to Massachusetts on a ship that first touched land in Maine. Historian Charles Mann wrote: "What Tisquantum saw on his return stunned him. From southern Maine to Narragansett Bay, the coast was empty—utterly void." What had once been a line of busy communities was now a mass of tumbledown homes and untended fields overrun by blackberries. Scattered among the houses and fields were skeletons bleached by the sun. Gradually Dermer's crew realized they were sailing along the border of a cemetery 200 miles long and 40 miles deep. Squanto's native Patuxet had been hit with special force; not a single person remained. At Massachusetts Bay, conflict and foreign diseases caused the Native American population to plummet from approximately 37,500 to 5,300. Around Delaware Bay, after some fourteen epidemics and several wars, their numbers fell from about 20,000 to about 4,000 by 1700. At Pamlico Sound, North Carolina, Native Americans were halved to 500 individuals after encountering Europeans. From then on, the fates of these fishes were in the hands of the white man.

Chapter 7

Providence and Plenitude

To live by a large river is to be kept in the heart of
things.
　　　　　　　—John Haines, "Moments and Journeys"

The cannon fired—but it was the *gunner* who was killed. His death in
1806 on the Falmouth, Massachusetts, village green was the only fatality
in the Coonamessett Herring War. The odd ammunition he loaded into
the weapon may have had something to do with it; the cannon's barrel was
packed with herring, as a statement by the "Anti-Herring Party." This
"fish war," begun in 1798, was one of several that were fought on a number of Atlantic rivers, where tensions between those who benefited from
free flows and those who profited from dams reached a fevered pitch. *War*
is a term used loosely; they were often guerrilla actions against dams, or
actual skirmishes between both sides, which later tended to terminate in
legal resolutions. They also were the inevitable product of an evolution of
the use of rivers from the sustainable harvests of Indians, to early colonists
for whom fish runs were manna from heaven, to millers and industrialists
who wanted to capture the potential energy of falling water with dams to
power machinery—and to hell with the fish.

The shift from Native American fishing practices to the colonial had drastically changed the concept of natural resources ownership. Indians had
organized within watersheds; colonists instead used rivers to mark the
limits of their lands. When the Massachusetts Bay Company's charter was
issued in 1628, the bounds of the grant stretched "betweene a great river

there commonlie called Monomack, alias Merriemack, and a certain other river there called Charles river." Rivers, of course, provided convenient and unequivocal boundaries, but the need for these markers was a colonial imperative. Indians had a far more flexible understanding of property rights. They never owned the land or the water they harvested in any definitive sense; what they claimed was the right to whatever might be obtained from that land or water.

The historian Ted Steinberg wrote that "[w]ith the British system of property relations as their guide, the colonists soon came to see the New England landscape as a vast bundle of commodities—of wood, fish, and furs." And, of course, the land itself. The opportunity presented by unsettled terrain that could be farmed drew colonists inexorably inland. Some came to central New York, near the source of the Susquehanna River, where fish came to offer salvation. The Susquehanna is a giant of an East Coast river, draining much of New York and Pennsylvania before emptying into the head of Chesapeake Bay. But it also is an odd one; the Native American name *Susquehanna* means "a mile wide and a foot deep." In 1785 William Cooper, a land speculator, set down stakes at the river's origin, the south end of Otsego Lake, to found the settlement that was to become Cooperstown. One spring a few years later, two hundred families there bordered on starvation; Cooper's heart was torn at seeing them "without a morsel of bread." In Cooper's words: "A singular event deemed sent by a good Providence to our relief; it was reported to me that unusual shoals of fish were seen moving into the clear waters of the Susquehanna"—what he described as herrings, but which almost certainly were American shad. Crude nets fashioned from twigs caught thousands; in less than two days, each family had an ample supply.

It's not clear how often shad and river herring ascended the 1,200 feet above sea level that far upriver, but it must not have been a rare occurrence based on the observations of James Fenimore Cooper. Cooper wrote in *The Pioneers* that "[e]normous shoals of herring were discovered to have wandered five hundred miles through the windings of the impetuous Susquehanna and the lake was alive with their numbers." Today the few shad that even attempt to swim up the Susquehanna stop below the first one or two dams, barely above tidewater—and nearly 500 miles shy of where they rescued Cooper's settlers.

Just one decade earlier and one large river to the north of the Susquehanna, American shad is known for an even more celebrated "rescue." The Schuylkill, a fourth-order river, enters the Delaware River near Philadelphia. Three-fourths of George Washington's Continental Army, about 12,000 troops, wintered at Valley Forge on the Schuylkill's right bank in the winter of 1777–78. The suffering they endured and the "miracle" of the shad run appear partly mythologized (something not unknown for George Washington's place in history). As McPhee pointed out, "it would not have been a leap of the imagination for him to anticipate the spring shad run and choose a campsite accordingly. He was a commercial shad fisherman." What is known for certain is that Washington wrote a series of diatribes to Congress, New York governor Clinton, and others that told of being encamped in an exhausted countryside in freezing, near-starvation conditions, and needing immediate assistance.

Yet the founding fish were on their way. The British, wishing to prevent a nutritional restoration of the enemy, tried to cut off the shad run by stretching a barrier net across the Schuylkill near Philadelphia. But the shad run did not coincide with the famine. Numerous accounts of their timely appearance are part of the "providentialist canard" that also colors accounts of the first Thanksgiving. McPhee recounts how historian Wayne Bodle found that the peak of the famine during icy conditions in early March was relieved by the arrival of supplies, well before shad had entered the river. Indeed, after a mid-March thaw, a new freeze left an inch of ice on the river on March 22. Shad likely did eventually reach Valley Forge in the course of their normal migration cycle, perhaps adding variety to the army's food store, but the notion that shad saved the Continental Army is a canard. And yet, if we discredit the shad's role at Valley Forge, we can still make a broad case for river herring, sturgeon, striped bass—as well as shad—as America's founding fishes.

Early colonists were woefully ill-equipped to actually harvest fish, despite knowledge that American waters abounded with them. In *The Unnatural History of the Sea,* Callum Roberts wrote: "The rivers they sailed from in Europe were by this time awash with human waste, choked with sediment, and, in their upper reaches, blocked by long chains of milldams and weirs. Not since the early Middle Ages had Europe's major rivers run cool and clear. By the late fifteenth and sixteenth centuries, the days when shimmering columns of fish fought their way upstream to spawn were long forgotten. In the rivers and estuaries of the New World, Europeans rediscovered what they had lost at home."

Rediscovery brought both amazement and unfamiliarity, the latter including incompetence when it came to actually catching the fish teeming before them. Not a single item of fishing gear was known to have been brought by colonists to Virginia. Being unprepared, but also in grave, if not desperate, need of fish as food led to almost comical improvisation. John Smith in 1608 on an exploratory trip up Chesapeake Bay wrote: "[W]e found . . . in diverse places, that abundance of fish lying so thicke with their heads above the water, as for want of nets, our barge driving amongst them we attempted to catch them with a frying pan; but we found it a bad instrument to catch fish with. Neither better fish, more plenty or variety, had any of us ever seene in any place, swimming in the water, then in the bay of Chesapeack, but there not to be caught with frying pans." On this same expedition the settlers caught some fish by "nailing them to the ground" with swords. The need for skilled fishermen was recognized with a 1610 memorandum from the Virginia Council to the authorities in London, asking for the next immigrants to include twenty fishermen and six net makers, and to select them with care. Fortunately, fishing practices soon evolved from cooking utensils and weaponry with what they learned from Indians and their own experimentation, likely fueled at times by hunger's quiet desperation.

Colonies, though, also were business enterprises. The English promoters of the Jamestown colony quickly protected their future investment in Virginia by securing a patent in 1606 from the King that granted them full possession of the fisheries. Despite the distance of the intervening ocean, this

right was taken seriously; by 1612 a law was in place that required strict tal-
lying and reporting of the production of all sturgeon and sturgeon products.
Fail to comply, lose your ears. Do it again, and you were condemned to one
year manning oars as a prisoner in the galley; a third time, and three years
in the galley.

These fish that in springtime magically appeared in rivers were
deeply appreciated, and *providence* was a commonly used noun for them.
Concerning the alewife in the Taunton River, an early observer wrote: "It
seems to be a sort of fish appropriated by Divine Providence to Americans
and most plentifully afforded to them so that remote towns [far upriver]
have barreled y'm up and preserved them all winter for their reliefe." But
over time the relationship between anadromous fish and European colonists
advanced along a complicated path, from a source of eminently reliable—
if not at times miraculous—sustenance, to one of the many contradictory
"services" that rivers could provide, services that would later force difficult
choices.

Yet when rivers still ran silver, astonishing landings were made
with primitive gear. A French priest in 1723 described fish ascending the
Kennebec "in such numbers that a person could fill fifty-thousand barrels
in a day, if he could endure the labor." And Kennebec records from the late
1700s and early 1800s describe an individual fisherman catching 500 salmon
in a season, four men landing 6,400 shad on a single day, and 1,000 striped
bass being taken in a single weir during a single tide. In *The Founding Fish,*
McPhee recounted an April 1770 newspaper report of a catch from the
mouth of the Hudson Estuary. One of the shad seines was so filled with
fish that the weight pressed it down, allowing great numbers to escape. A
second net was then thrown around the fish, a third around the second, and
a fourth around the third. "The number of shad that were taken by the first
net was three thousand; by the second, three thousand; by the third, four
thousand; and by the fourth, fifteen hundred; in all, eleven thousand five
hundred!"

Fish also helped to support colonial agriculture. Early on, a river running silver was not only a source of food but also of fertilizer. "It is an
astonishing sight to paddle down the Restigouche and see the farmers

A large catch of alewives being landed in a haul seine at Sutton Beach, Albemarle Sound, North Carolina, 1887
NOAA National Marine Fisheries Service

'smelting'—scooping up the little fish in handnets. The amount they take is incredible, and most of their potatoes spring from this fishy manure." And pure waste in the midst of plenty was not unknown. An account from the Chesapeake in 1777: "Saw a seine drawn for herrings and caught upwards of 40,000 with about 300 shad fish. The shads they use but the herrings are left upon the shore useless for want of salt. Such immense quantities of this fish [are] left upon the shore to rot; I am surprised it does not bring some epidemic disorder to the inhabitants by the nauseous stench arising from such a mass of putrefaction." Such misuse was likely anomalous and due to the occasional problem of obtaining sufficient salt; however, most families in the Susquehanna region around 1827 annually "put down" 100 to 300 leftover shad into their fields, and this practice continued for decades.

Acclimation of Europeans to make use of American resources such as diadromous fishes took some time. But once the settlers had adapted to the New World well enough to provide for themselves, the inclusion of fish with winter livestock, deer, small game, waterfowl, corn, and beans was a healthier and more balanced diet than the grain-based diets they left behind in Europe. In fact, the proportion of fish in their meals declined as time went on. On the lower Potomac, archaeologists found that for sites occupied between 1700 and 1750, fish made up only 1.5 percent of the bone samples, compared with about one-third for the seventeenth-century sites. What fish colonists did capture were with a mix of Native American and European techniques, filtered through their own circumstances.

Haul seining, a highly effective Indian practice, was refined by the colonists. In its early uses by the Europeans, a vertical wall of net was arced through a river pool and then "hauled" back to shore using raw manpower, raw animal power, or a windlass-style winch. In the beginning these efforts were communal, with settlers receiving a share of the catch. Later the seines became longer and longer, and the fisheries became professionalized as businesses with paid laborers. Even with these technical refinements and specialization, there still was some community involvement. One example is the "widow's haul." For years following the July 1778 Revolutionary War Battle of Wyoming between settlers and the Iroquois, the complete catch of the first Sunday of the fishing season was given to the widows and orphans of the Wyoming Valley, Pennsylvania.

By 1827 the Stump Farm operation at Susquehanna Flats was the most prodigious haul-seine fishery at the time, the paragon for any species. Willis later described an extraordinary catch made there: "In the spring of 1827 Thomas Stump owned and operated the largest shad fishery in the United States, immediately below the railroad bridge on the opposite side from Havre de Grace. At the mouth of the river, his seine was laid across the river and down for miles along the shore below the village of Havre de Grace. A violent wind commenced, which put a stop for four days and nights to any further

action with the seine, the wind blowing constantly down the river, and no shad
could get past the seine. The wind at one o'clock on the fourth night changed
and blew directly up the river, and by daylight the outer end of the seine had
reached the windlass. . . . At eight o'clock one hundred wagons and carts that
had congregated from Lancaster and Chester Counties were loading shad at
$4 per hundred. . . . It took three and one-half days to get the seine inshore;
hundreds of wagon and cart loads of fish were put on lands as fertilizers from
that one haul." Though no systematic estimate was made, this single haul was
thought to have landed an astounding 15 million shad and river herring.

Further development of haul seining included development of "shad
batteries." Batteries were places amenable to haul seining where debris was
cleaned out and everything was fitted to catching and packing shad. This
was taken to its apex with the construction in the early 1800s of Battery
Island, a man-made islet on the Susquehanna Flats strategically sited and
designed for efficient removal of shad from Upper Chesapeake Bay. A survey in 1836 showed that what was then known as Donahoo's Battery had
reached more than an acre in size and provided a hauling ramp, housing for
workers, and facilities for cutting, salting, and packing fish.

Whereas Shad Battery Island privatized a portion of the Bay, temporary "shad floats" were used to circumnavigate riparian rights. Shad floats
provided a landing area for haul seines right out in open water. These were
ambitious enterprises, 75 to 100 feet wide and 200 to 300 feet long, made of
logs with rough flooring, and accommodating dozens of men in a one-story
building, a kitchen, a mess hall, stables for the windlass horses, storehouses,
and salting sheds, plus an outside apron platform for landing and sorting
the catch. Beneath the buildings were vats to hold fish awaiting transport
to shore; catches were so large that these measured 16 to 18 feet long, 4 feet
wide, and more than 3 feet deep. So successful were shad floats that they
were used in the lower Susquehanna River for about a century.

After Squanto in 1621 taught the Pilgrims how to catch eels, eel fishing
by the colonists took off in myriad forms. Today it is nearly impossible

to interest an American in eating eel unless that person is originally from an Eastern European or Asian nation, or if the fish is smoked or hidden inside a sushi roll. But eels rapidly took on much of the same importance to the settlers as they had to Indians. In Quebec pioneers were granted lands with direct access to the St. Lawrence River and its tributaries such that eel weirs could be guaranteed placement, providing a significant portion of settlers' protein consumption throughout the year. Before the Welland Canal was built linking Lake Ontario and the St. Lawrence to the other Great Lakes, eels aggregated at the farthest barrier to upstream migration in Lake Ontario—Niagara Falls—in astounding numbers. Nineteenth-century naturalists at the base of the falls commented that "hundreds of wagonloads . . . would hardly be considered excessive by those who have visited the spot at a suitable time of year." Much as Indians had used eels creatively for non-food needs, colonists applied eel oil to soften harnesses and for medicinal uses, and eel skins were preserved for lining horsewhips.

The pronounced efforts by early colonists didn't take long for even their rudimentary fisheries to make observable dents in some anadromous fish populations, with the first concerns being raised in Maryland within thirty years of Captain John Smith's explorations of Upper Chesapeake Bay. In 1638 the General Assembly of Maryland restricted netting in a tributary to the Potomac, named Herring Creek. Over time for many rivers, well into the twentieth century, "lift periods" were specified in which all nets were required to be out of the water. For instance, in 1807 seining was prohibited on the Susquehanna between six p.m. Saturday and five a.m. on Monday. These hiatuses provided the opportunity for some fish to reach the spawning grounds and for the fishermen to rest at ease with the knowledge that no one else was out there harvesting fish when they weren't.

Another level of protection for the fish was offered in 1735 with "An Act to Prevent the Destruction of the Fish called Alewives," passed by the Massachusetts Bay Colony, which included the statement: "That no dam shall, hereafter, be erected across any river or stream, thro' which alewives or other fish have been accustomed to pass into ponds, in which there is not made and left a convenient sluice or passage for such fish." Such laws, however, were rarely enforced, which led to frustration by the herring advocates. Society's equilibrium did shift briefly toward effective passage during the Revolutionary War. The British naval blockade had shut off the supply of cod and other ocean fish to Maine's coastal towns, and the value of anadromous fish as an alternative was apparent. Suddenly, many towns responded by demanding legal action against milldam owners who had long violated existing laws by blocking runs of fish in their local rivers. Petitioners in August 1776 presented an argument for fish passage on the Presumpscot River, noting the once-plentiful supply of "Salmon, Alewives, Shad & Other Sorts of Fish that frequented the said River. . . . And our fishing on the Banks as well as on our Coast off Shore being in a great measure impracticable by reason of the Enemy's cruisers that infest our Coast, reduces us to the necessity of Adopting some method whereby the Fish may come to us."

Growing competition for flowing water to provide either fish or mill power led to more and more conflicts, with the Coonamessett Herring War only one of the more colorful. Before Coonamessett there was Atkinson's Mill Dam Raid in 1732. Stephen Atkinson, who had arrived in Pennsylvania with the eponymous William Penn, built the first water-powered mill in Lancaster County, in Conestoga Creek, between 1714 and 1728. By 1732 residents upriver of the barrier were so angry over the loss of the shad run that one night they sent a raiding party to destroy the dam. But Atkinson rebuilt the barrier, and it operated for decades more. In this and many similar situations along the Eastern Seaboard, the courts attempted to settle the disputes by allowing the dams but also specifying fish-passage modifications to be made. In Atkinson's case he did offer in 1731 to construct a twenty-foot-wide fish passage alongside his dam, but he was not willing to pay for it himself. No public funds were forthcoming, and so, despite his commodification of the commons, he did not apply any of his profits to remedying the blockage. Indeed, in Pennsylvania and elsewhere, laws were passed that specified that fish passage be incorporated at existing or newly built dams, but these requirements were often ignored, with little enforcement.

Some fish wars did not involve dams, just competition for fish. A simple weapon on the Susquehanna was "shingling," where a wooden or tin shingle was fastened to a short cord and anchor and tossed in a strategic location, such as a fishway, to swing in the current and frighten shad from moving upriver. In the 1850s Great Safe Harbor Shad War, seine fishermen from Columbia, Pennsylvania, traveled downstream to destroy the Safe Harbor fishermen's nets, stone traps, and weirs. The invaders were driven off with rocks and long boat poles and, some said, the threat of cannon fire. Another skirmish occurred in 1860 when fishermen, again from Columbia, tried to seize an island that was a good fishing location. This time the defenders beat the assailants so savagely with sticks and stones that the defenders were found guilty of assault and battery and fined five dollars each. The following year a battle involving another island resulted in a killing with a bullet fired by a fisherman's son. The father, who ordered the boy to fire, was sentenced to eleven and a half years in Eastern State Penitentiary.

More often dam owners won contests about dam removal or the installation of costly or water-wasting (from their point of view) fish passage by waiting for the fish runs to decline and wearying their frustrated opponents. But some fought back more directly, claiming that to yield water power to fish would hurt the local economy, an argument termed "pickerel vs. payroll." And in 1791 one mill operator on Cobbosseecontee Stream, Robert Gardiner, retaliated against the annual protestations, town meeting actions, and fish committees by arguing that the Massachusetts Legislature should forever exempt his river from a newly strengthened law compelling the installation of fish passage by simply denying that salmon, shad, and alewives had ever ventured up that system—historical ecology in reverse. A number of local residents filed affidavits to the contrary, which caused the Legislature to reject Gardiner's request, but they did not force him to actually comply with the law, and town minutes over the next five years indicate annual attempts that also were ignored. Gardiner not only persevered in his noncompliance—he triumphed. Later records are inadequate, but it is known that in 1806 Gardiner won the exemption he had requested, and that he and his investors quickly added nine more dams to his original two along the Cobbosseecontee.

Henry David Thoreau's first night spent outdoors in his entire life, accompanied by his brother John, was on a spit of land on the Concord River, upstream of the town of Billerica. As they rowed downriver that afternoon, Henry Thoreau delighted in the diversity of the river's wildlife and the occasional solitary fisherman who lingered by its shores. But when the brothers neared Billerica, a sudden shift in mood occurred; Henry Thoreau knew that the salmon, shad, and alewives were missing from the river because the dam had ended their migrations.

Thoreau saw the problems the dam presented beyond their blocking fish migrations. He wrote: "At length it would seem that the interests, not of the fishes only, but of the men of Wayland, of Sudbury, of Concord, demand the levelling of that dam." The dam had not only halted the harvesting of fish for food and fertilizer, but it had also flooded the meadows for many miles upstream. And in the heavily wooded Massachusetts countryside of that time, natural meadows were essential to farmers for hay to feed their livestock. The farmers now stood idly "with scythes whet," vainly "waiting the subsiding of the waters." "So many sources of wealth inaccessible," Thoreau exclaimed.

The first dam at Billerica, constructed in 1710, supplied power for a small woolen mill, while later its millpond provided water for the Middlesex Canal, one of America's first large-scale engineering projects. Henry and John traveled through the canal to the Merrimack instead of farther down the Concord itself because the canal diversion left little water for the river in August. Only one year after its completion, the dam faced its first lawsuit. This was followed by a litany of controversy: In 1721 the dam was removed at the order of the court; in 1722 it was rebuilt; in 1723 the dam was removed by an angry band of farmers; in 1791 a fishway was added; in 1798 the dam was again rebuilt; around 1800 the dam was raised for the new canal system; in 1809 and 1815, legal efforts to remove the dam failed; in 1829 a new dam was constructed against the old dam; in 1859 the dam was ordered removed; and in 1861 the dam owners appealed and lost, but then the Civil War began and efforts to remove the dam ceased. In 1859 a professional surveyor helped to gather evidence for the defendants looking to remove the dam. The surveyor was Henry David Thoreau. And here is an irony: The dam was recognized and protected as historically significant in 1938—partly because of the very efforts of Thoreau to take it down.

As the Industrial Revolution progressed, the declines that Thoreau noted were seen in rivers up and down the Eastern Seaboard. One observer, Thaddeus Norris, wrote in 1868: "Shad, at one time, entered every river on our coast which furnished the requisite spawning-beds, and ascended until some barrier opposed their course; every tributary was crowded with them. Civilization, and its attendant enterprise, prosecuted without provision for the passage of the fish to and from their spawning-grounds, have driven them entirely from some rivers, and lessened their numbers materially in others, that shad are now considered rather a luxury, than one of the chief staples of life, in their season."

Thoreau's interest was especially attracted to the plight of the shad, which migrated each year up the river only to be met "by the Corporation with its dam":

> Poor shad! where is thy redress? When Nature gave thee instinct, gave she thee the heart to bear thy fate? Still wandering the sea in thy scaly armor to inquire humbly at the mouths of rivers if man has perchance left them free for thee to enter. . . . Armed with no sword, . . . but mere shad, armed only with innocence and a just cause . . . I for one am with thee, and who knows what may avail a crow-bar against that Billerica dam?

The Billerica Dam is still in need of that crowbar. The Concord's salmon and shad are nothing but ghost fishes. In 2000 a small alewife restoration began there. In 2012 the website for this program had an asterisked statement on the bottom of the page. It read: ***This program is currently on hiatus due to a moratorium on fish transfers and low fish counts; however, we would still like to know if you are interested.**

Chapter 8

Floating Caskets
and the Pennsylvania Navy

Progress celebrates Pyrrhic victories over nature.
—*Karl Krauss (1874–1936)*

"I feel a lot of personal responsibility for the dam," said Amy Roe as we sat on the west bank of the Susquehanna below the giant Conowingo. This responsibility was genealogical, though, not direct; her great-uncle Charlie had helped to engineer it, and he'd done a good job—too good, in her opinion.

I'd met Roe at the Conowingo Dam because I wanted to know first-hand the passion that had prompted her to write a 900-page PhD dissertation on the history of American shad in the Susquehanna. Titled *Swimming the Occidental Current: A Resource Hermeneutics of Fishery Collapse,* it is a voyage through time down the sorry course of the Northeast's longest river, a fantastic factory for shad before the "occidental current," her metaphor for the relentless commodification of the river that murdered its primordial bounty. Hers is a rich and idiosyncratic work, blending jargonish state-of-the-art notions of social science, myriad details about the fish and the river, and even her observations of primitive fishers in Peru—an ambitious synthesis arising from a distinctive personality. Roe wrote her thesis in pencil, and she tells me excitedly about the tactile pleasure of working with a single sheet of paper at a time on a wooden drafting board and an Alvin Draft/Matic mechanical pencil holder in which she varied the lead density according to the humidity of the day.

It's late April 2012, normally a time of high flows, but the Susquehanna is shockingly low because of a rare spring drought across the Eastern

Seaboard that followed an almost snowless winter. Nonetheless, the pool below the dam is alive with fish and fishermen, and birds and birdwatchers. A birder with a telescope points out bald eagles on a tower on an island, among numerous cormorants, vultures, herons, and gulls that flew by or perched on rocks and trees, studying the water. Two anglers in a boat land a couple of shad, and many fish swirl on the surface. The birds and the people are concentrated here because the fish are, too; the dam has stopped all upriver migration. I jump up and break Roe off as she speaks to point to a bald eagle right in front of us, making a stoop dive out of the sky to snatch a shad that ventured too high in the flow. The river is too low to pass shad over the dam on the fish elevator, so what is typically only a tiny relict run has aggregated, providing a false appearance of plenitude.

Landing shad in Upper Chesapeake Bay, 1905; photo that inspired Amy Roe.
Historical Society of Cecil County, Maryland

Roe's ancestral home is Perryville, at the mouth of the Susquehanna. This landscape remains at her core; she says that when crossing the nearby Tydings Memorial Bridge she feels strongly that she is at the Center of the Universe. Shad fishing was part of her lineage. Her ancestors had fished shad, shad being so dominant a quarry that they were known simply as "fish." Although the river was a major part of her childhood and she knew of the importance of shad to her family and the region, it wasn't until years later that she had an epiphany about the magnitude of changes that had occurred in her world. It struck her when she picked up a book on the history of Upper Chesapeake Bay and was captivated by one particular photo, one she still occasionally stares at till this day. It shows men standing in the water landing a net laden with shad, in numbers orders of magnitude above what she had experienced in her time along the river. In her own deeply probing style, she needed to know how, in less than a century, the ability of the Susquehanna to provide like this had been squandered. It was an intellectual journey that took years.

The Susquehanna is the most sprawling watershed on the East Coast, draining a territory of 27,510 square miles across New York, Pennsylvania, and Maryland. But despite its size, it's oddly non-navigable, the largest river in the United States that can't support ship traffic. Though now thoroughly subdued, it once was wild, "one of the most furious, perilous, and ungovernable torrents in the world." Though ships couldn't handle the shallows, American shad could, with many millions running silver more than four hundred miles from Havre de Grace at its mouth to Lake Otsego.

The fish's presence was highly anticipated: "The luscious shad again came up in countless myriads, inviting the toilworn emigrants from the dangers of the field, to the sports of the stream, from the half famished abstinence of the camp, to feast on the richest of nature's dainties. Hope, and joy, and confidence began to prevail." In a letter from 1881, Gilbert Fowler, who was born in 1792, recalled so many shad in his youth in Bloomsburg, Pennsylvania, that their migration caused a spectacle viewed in its approach from quite a distance. At the Webb Fishery, west of Scranton, someone with a spyglass sat atop a vantage point and scanned the river constantly, looking for signs of approaching shad. So large were the schools that "[f]rom the banks of the river at this fishery could be seen great schools of shad coming up the river when they were a quarter of a mile distant. They came in such immense numbers and so compact as to cause or produce a wave or rising of the water in the middle of the river extending from shore to shore."

As the population of the Susquehanna Valley grew and more land was cleared for farming and for lumber, there was a growing need for water power to drive grain mills and sawmills. This led to the placement of more and more dams on tributaries, with some streams becoming lined with mills like beads on a necklace. In Pennsylvania's Lancaster County alone, 147 milldams were constructed between 1756 and 1776; this leapt to 332 between 1805 and 1817. Though destructive to the shad runs that went up these tributaries, shad still traveled up the main-stem Susquehanna where local provisioning could still occur.

The commodification of the main-stem Susquehanna began in earnest with the development of a canal system. Early efforts to make the Susquehanna navigable failed, and this watershed, like many rivers prior to the development of alternatives such as railroads and highways, had a series of channels carved into its surrounding landscape. A 1-mile-long canal completed in 1797 enabled boats to bypass the Conewago Falls. A 9-mile canal along the lowermost river opened in 1809, but it was dangerously fast and narrow. In the 1830s the 43-mile-long Susquehanna and Tide Water Canal was built to bypass the rocky shallows of the lower river. By 1856 an extensive network developed, eleven canals totaling 634 miles, including connections to the Delaware River and another deep into New York to the Erie Canal, and yet another linked to the Ohio River via railroads.

These canals required reliable sources of water, which spurred the building for the first time of dams on the main stem, or on major branches, that blocked long stretches of river. The 900-foot-wide Nanticoke Dam, completed in 1830 on the Susquehanna's north branch, just south of Wilkes-Barre, Pennsylvania, ended long-standing riverbank seine fisheries that had sustained families for generations. When its final elements were put in place, so ended a shad run that had reached central New York for millennia. The Columbia Dam, built only five years later, shortened the unshackled portion of the river to just forty-three miles from its mouth. Neither dam had been built with a fish-passage facility, though the Columbia had a shallow sluice to allow rafts through that did not serve to pass shad. Fishermen upriver of the Columbia Dam were so angry that in 1838, the Pennsylvania Legislature approved legislation requiring its modification so that one hundred feet of the dam was open to migrating fish. But the dam owners simply ignored this edict, as they did when follow-up legislation commanded it again in 1851.

In 1866 some 400 to 500 fed-up fishermen's delegates converged in Harrisburg, the state capital. The Legislature responded by requiring fish passage within six months on major Susquehanna dams. A retrofitted forty-foot-wide fix at Columbia failed to pass shad, as did additional legislated modifications in 1873, 1880, and 1886. In 1881 a rare voice of reason was ignored; in consideration of a proposal to dredge the upper Susquehanna River for steamship navigation, Colonel William Ludlow of the US Army Corps of Engineers advised against this measure, and even went so far as to recommend dam removal: "[I]f the carrying traffic now done by the canals was transferred to the railroads, and the dams destroyed, the supply of fish resulting along the river would add more to the wealth and comfort of the people than is now produced by the canals." It's not surprising that it wasn't the technical adjustments, but instead breaks in the Columbia Dam in 1873, 1877, and 1895 that provided effective, though temporary, passage. The dam finally broke in 1896 and was not repaired because canal navigation had concluded two years earlier, temporarily returning forty miles of river to the fish.

The Susquehanna's water quality also declined. Farming and logging activities led to silt-laden runoff, much of which became trapped behind the many dams. But some of the eroded materials reached the river mouth, adding two to three yards of sediment between 1760 and 1860. Industrial mills and tanneries dumped contaminants into the river, and municipal discharges added human waste. Farmers, no longer able to catch shad in their local streams, needed to travel long distances to buy shad from the large commercial fisheries in what unobstructed main stem remained.

Anthracite coal from the North Branch, a ready form of convenient energy, supplanted water power once people figured how to actually burn the hard "stone coal." Canal traffic shifted from agricultural and water-powered manufactured products to the growing appetite for coal for heating homes and firing iron furnaces. But the iron mining itself poisoned sections of the river as the toxic waters that accumulated in the shafts were pumped into the watershed. To this was added vast amounts of coal silt from the breaking and cleaning of the coal once brought to the surface, which also resulted in turning the waters acid. The scale of coal wastes was so huge that an industry referred to oxymoronically as the "Pennsylvania Navy" developed to scavenge coal particles from the bottom of the river for household use. So much coal silt accumulated behind the Holtwood Dam that a coal-fired generator was installed, making this the only hydropower facility in the world that also burned coal to make electricity. The occidental current was indeed pushing hard against the shad.

Edison's late-nineteenth-century invention of the electric lightbulb increased demand for power both day and now at night and led to the building of more, larger dams for hydropower, or "white coal," as river water was known. Four great dams were constructed in the lower reaches of the Susquehanna between 1910 and 1928; all still provide hydropower today. The first two dams, York Haven and Holtwood, were built with fish passage that utterly failed. Recognizing this, the builders of the third dam, Conowingo, only ten miles from the river's mouth, did not include passage facilities, which made passage upriver on the other dams moot. Therefore, when Safe Harbor Dam was squeezed in upriver of Conowingo, between Holtwood and York Haven, it made no sense to provide passage there either. As a sort of penance for destroying the shad run, dam owners paid Maryland and Pennsylvania an annual fee to stock fish and amphibians in the river above the Safe Harbor Dam. The list of creatures actually stocked is bizarre. They did not include any shad. But between 1932 and 1946 they did include 89,500 tadpoles, 159,300 frogs, 185,820 minnows, and 538,761 suckers. How this represented any form of compensation for the loss of shad is not apparent.

Soon after World War II, Pennsylvania launched a study of its loss of migratory fishes and the failures of fish passage. Part of this investigation included trucking spawning-ready shad past dams for stocking upriver. But no young shad were ever found to have resulted from this. Despite the lack of evidence for success of the trucking approach, an experimental trap was fitted to Conowingo in 1972 to capture fish for trucking, and this was expanded in 1982 to a program that moved as many as 16,000 shad in a year upstream. In 1988, however, the dam owners agreed to build a fish elevator with the potential to move as many as 750,000 shad over the Conowingo in a single season. But where would all these fish come from?

Virtually all Atlantic rivers of any consequence were commodified during the Industrial Era; the Susquehanna was just one of the most pronounced and well-documented cases. A smaller river farther north also endured the countless slashes of "progress." The Blackstone winds its way south for forty-eight miles from Worcester, Massachusetts, emptying into Narragansett Bay at Providence, Rhode Island. Though only of modest size, the Blackstone drops forcefully at a fairly steady average of nearly ten feet per mile, which made it ripe for mill power. Because of its intense industrialization, in the late 1800s it was called with some admiration "America's hardest-working river." Of course, industrialization has nasty environmental consequences, and for this in 1972 it was named with some shame as the river in America most polluted with toxic sediments. Like so many rivers of the Eastern Seaboard, it can recall a goodly portion of the elemental historical arc, from sustained Native American fisheries, to providential source of fish for European colonists, to focused industrialization, to a more-amorphous post-industrialized existence. It's just that for the Blackstone, the arc is more curved and more complete than for many other rivers.

The Blackstone's natural falls at Pawtucket were a hindrance, but not an obstacle, to migratory fish, assisted by "the Indians at first, and the white men after them, [who] took much pains to smooth them off by battering down the projectile points, that the fish, in time of running, might the more easily ascend them, which feat they actually performed." Indians were said to have stood on the rocks below Pawtucket Falls and, with bows and arrows, shot salmon as they attempted to leap the falls. Sufficient salmon and shad passed this obstacle, or used a small side channel called Little River around it, that major seasonal Native American encampments existed well into Massachusetts. Later, salmon formed the "chief article of diet" for farm laborers in the Blackstone Valley. The fishing was easy in the pool below Pawtucket Falls; according to one observer, "All we had to do was sweep our nets when we saw the breaks in the water by the on-rushing fish and intercept them before they could get to the falls."

A series of projects altered the physical river, beginning with a bridge built in Pawtucket in 1713 that narrowed the main river and blocked the side channel. In 1718 small, partial dams were replaced with a bank-to-bank dam at the Pawtucket Falls, one which had the usual effect. The structure was recognized years later in a petition to the Rhode Island General Assembly as being erected in such a manner as "to prevent the fish passing up said Falls whereby a valuable fishery was much Injured in said River." As a result of these changes and the blockage of the fish runs, Little River was reopened by William Sargent as Sargent's Trench. Depending on the account, opinions on the Trench ranged from "it wholly failed in its purpose" to "this arrangement gave a full supply of fish to the people above."

The remainder of the eighteenth century was a seesaw battle between the agrarian forces that wanted to harvest abundant fish runs upriver and manufacturing interests that sought to use all the Blackstone's waters for mill power. A dam was built across Sargent's Trench that blocked upriver movement completely, but in 1741 a Grant of Privilege was made to residents to build another trench, which worked well until a second dam on Sargent's Trench blocked passage through the new trench. Once again, a new trench was built, this time funded through a lottery conducted in 1761. And in 1773 an act was passed by the legislature, "making it lawful to break down and blow up rocks at the Pawtucket Falls, to let fish pass up."

But from then on, the pressures of industry dominated. The act in 1773, which had the potential to disrupt mill operations, was repealed only one year later. In 1790 the process of building a second dam higher on the falls was launched by Samuel Slater to power his cotton-spinning mill. Yet there was resistance; infuriated fishermen, artisans, and farmers filed lawsuits to

stop it. In 1792 four frustrated citizens took it further, destroying the partially built dam—one which would have lacked fishways—because it was
erected before the courts had ruled on the case. In 1805 commercial fishermen requested their limitations on fishing below the first dam be repealed,
and soon the sparse remaining stocks were fished out, salmon runs ending
first, followed by shad, and then by river herring.

During the 1800s anadromous fish had little standing concerning use
of the river. Mill owners argued with disdain that their need for power was
more important than "a trifling shad and alewife fishery [that] does not pay
for the grog expended in taking the fish," and that "leaving dams open for
migratory fish would only accommodate a few individuals with an opportunity of spending time in fruitless pursuit after a few scattered Herring."
The Industrial Revolution ramped up quickly on the hardworking river. By
1830 there was one dam for each mile of the Blackstone and every tributary
had been dammed; eventually this little watershed was dissected by as many
as two hundred dams.

During the 1900s the Blackstone River was a depository for a toxic stew
of dyes, sewage, heavy metals, detergents, and other contaminants. A high-
light of this era occurred in 1954 when the rain from Hurricanes Carol and
Diane caused the over-dammed river to wash out the Woonsocket cemetery, with residents watching caskets float downstream. Over more than a
decade after the Clean Water Act, the river was still in hideous shape; one
observer in 1984 said: "It was brown and black and full of suds and foam
. . . [a]nd you couldn't get close to it because it stunk so bad." But relief from
two centuries of unfettered abuse was around the corner, with many people
getting close to the Blackstone again, both physically and metaphorically.

Over the course of the Industrial Revolution, the characters of the
Susquehanna, the Blackstone, their tributaries, and the whole (or parts) of
virtually every Atlantic River passed from integral, free-flowing elements of
the region's primordial ecology to the sundered and captive course of Roe's
occidental current. When Roe defended her dissertation to her academic

committee and walked the group through the Susquehanna's sorry history, she had a side conference with one of its members. He told her, "Amy, write it off. Devote your efforts to other rivers that can still be saved." Devastated by the truth in that statement but unable to let go of the center of her universe, Amy Roe went home and bawled and bawled.

Chapter 9

Billions of Fish in Hot Water

It has been said that streams are the gutters down which
flow the ruins of continents.
—*Luna Leopold et al. (1964)*

Atlantic Coast rivers are fettered and tired, falling to the sea with the stilted
tempo of the subjugated. The Industrial Revolution strangled these sinuous
and dynamic waterways, turning many of them into rigid single-channel
sluiceways that stop and start as they pond and then pour over dams.
Ironically, there is a lingering romantic strain for the migratory fish runs
they choked off long ago that echoes through the names that developers
sometimes choose for residential complexes. This pull toward yore in the
naming of the spanking new provides a false veneer of timelessness. Witness
their sometimes comical attempts to render freshly minted McMansion or
townhouse spreads with the flavor of antiquity; they can't seem to inject
enough extraneous vowels. Someday we may feel nostalgic about Olde
Harbour Pointe Towne—but not yet. Likewise, the cookie-cutter "Shad
Row, "Salmon Run," and "Herring Brook" developments.

But just as this nomenclatorial jujitsu fails to make something genuine
out of nothingness, many of the fish-based place names that linger today in
fishless places do so as honest echoes of plenty from long ago. It is sadden-
ing to be reminded of what has been lost, but rivers and streams, villages
and towns, and other geographic features carry memories from both Indian
and colonial times. A small sampling, from Maine: *Passagassawakeag*—
place where we speared sturgeon by torchlight; *Androscoggin*—the high
fish place; *Skowhegan*—spearing station; *Cobbosseecontee*—the place where
sturgeon abound. Rhode Island: *Misquamicut*—a place for taking salmon.
Connecticut: *Naugatuck*—lone tree by the fishing place. New York:

Shekomeko—place of eels. North Carolina: *Chocowinity*—fish from many waters. Fish Trap, Georgia, is named after an Indian weir. Indeed, Georgia has the largest percentage of Indian place names of any US state. Of its fourteen major rivers, twelve have Indian names today.

Massachusetts is a hotbed of colonial diadromous fish-based monikers, with no less than five Herring Brooks on Cape Cod alone, a Herring River, an Alewife Brook, an Eel River, and two Bass Rivers, but also eleven Mill Rivers. New Hampshire's Ammonoosuc has a deep pool where it meets the Connecticut River called the Salmon Hole; Native Americans took quantities of salmon there up to twenty-five pounds. In New York there is a hole where a large Hudson tributary, the Rondout, receives the also considerable flow of the Wallkill. Known as the Sturgeon Pool, is there any other reasonable explanation for that name other than sturgeon once having been found there before a presently useless dam blocked them from it?

Even Charles Dickens spoke against the travails of an anadromous fish in the Industrial Era. In the July 20, 1861, issue of the weekly magazine *All Year Round,* he wrote: "The cry of 'Salmon in Danger!' is now resounding throughout the length and breadth of the land. A few years, a little more over-population, a few more tons of factory poisons, a few fresh poaching devices . . . and the salmon will be gone—he will be extinct. Shall we not step in between wonton destruction . . . and so ward off the obloquy which will be attached to our age when the historians of the nineteen-sixties will be forced to record that: 'The inhabitants of the last century destroyed the salmon.'"

Today the postindustrial river lies wounded, but recovering. And it's not even completely "post"; it's just that rivers are no longer so thoroughly dominated by manufacturing plants. Give considerable credit to the Clean Water Act of 1972, a powerful piece of legislation that turned the tide of pollution in many water bodies around the United States, and that continues to ripple toward environmental recovery today. And to the environmental movement that helped spawn the Clean Water Act, and to

Cleveland's Cuyahoga River catching fire in 1969, which helped to inspire the environmental movement and the Clean Water Act, and to enormously influential texts like Rachel Carson's *Silent Spring* that were fundamental to Americans caring about the physical, chemical, and biological states of their nation.

Though the trajectory of major East Coast rivers has been toward environmental recovery, challenging legacies remain from the Industrial Era: Dams that no longer serve useful purposes are the most obvious and least addressed. But contaminants that were dumped into rivers following the adage that "the answer to pollution is dilution" still linger and wield their effects. In Maine noxious fumes from the Kennebec and Androscoggin Rivers once peeled the paint off buildings; today's sediments still contain dioxin from the textile and paper mills that dumped pulp and other wastes into their waters. The Androscoggin has never met Clean Water Act standards. Even now, near some mills on the river, oxygen has to be pumped in during the warm summer months so that algal blooms do not kill fish. "Hot spots" of polychlorinated biphenyls (PCBs) used by General Electric in the manufacture of electrical transformers and discharged into the upper Hudson River still lie in sediments where they leach out, compromising the edibility of the river's abundant fish life. But PCBs are at the forefront of a litany of contaminants, such as dioxin and chlordane and heavy metals like lead and mercury that permeate the Hudson Estuary. PCBs are definitely *not* one of the good things GE brought to life, but GE is dredging some of these hot spots. Still, PCBs are found throughout the system almost everywhere else, too—even in the Arctic's polar bears.

The Delaware River, awesomely industrialized near Trenton, New Jersey, was once so thoroughly polluted that the area functioned as a "chemical dam": When water temperatures warmed in spring, low oxygen levels and toxins made it impassable for migrating fish. Just to the south, the giant Chesapeake Bay, "America's Estuary," is being overwhelmed with nitrogen runoff from the swelling population centers and farms that surround it. This has created what David Secor calls a "temperature-oxygen" squeeze for sturgeon and other fishes that occurs in large sections of the Bay. Surface waters warm in summer, driving fish toward the bottom, but bottom waters become depleted of oxygen, driving fish toward the surface. Hence, the "squeeze"—where is a fish to go?

Hopefully, not to North Carolina, where huge open-air hog waste lagoons, often as big as several football fields, are prone to leaks and spills. In 1995 an eight-acre waste lagoon in North Carolina burst, spilling 25 million gallons of manure into the New River. The spill killed about 10 million fish and closed 364,000 acres of coastal wetlands to shellfishing. Nor to the Savannah River, which forms most of the South Carolina–Georgia border. As of 2010 it had the fourth-highest annual discharge rate of toxic chemicals of American rivers, a cocktail of almost 10 million poisonous pounds. Nor as far as the St. Johns River, Florida, the lower reaches of which regularly have algal blooms, fish kills, and fish with deformations and lesions.

Many not-quite-postindustrial rivers also still have major cargo and oil terminals. Surprisingly, the comings and goings of large tankers and other ships may be retarding sturgeon recoveries. The Delaware alone sees some 3,000 round-trips each year. With the extraordinary length and depth of these vessels today leaving little space between the bottom of their hulls and the dredged channel bottoms, combined with the great hydrodynamic forces generated by the vessels and their huge propellers, a biologically significant number of Atlantic sturgeon are struck, washing up dead on the river's shores, often severed through the torso or head. This is also being seen on Virginia's James River and Maine's Kennebec.

The rivers of today may also be home to an electric generation or "power plant," or two or three, or many more, in some cases. In fact, seventy-nine are found on US Atlantic rivers and estuaries. Most power plants are sprawling installations that generate anywhere from a village to a small city's worth of electricity, usually by burning coal or natural gas, or by splitting atoms. To mitigate all the resultant heat, traditional power plants need to run vast amounts of river waters through their machinery.

The Hudson River Estuary is power plant central for central power plants, its waters providing a wet chill to more than a dozen facilities. Today nine are of the older "once-through" cooling design, two are "wet closed-cycle" that lose modest amounts of water to evaporation as it is recycled, and three are state-of-the-art "dry closed-cycle." The maximum amount of water that can be withdrawn by the once through–cycle plants is gargantuan; for Indian Point, more than 2.4 billion gallons per day. And they are hot, too; the highest permitted temperature in its discharge plume is 110 degrees Fahrenheit, which is above the human pain threshold. Together, the Hudson power plants pass about half of the river's normal freshwater flow through their pipes daily.

Though central power plants have existed for more than a century, their effects on aquatic environments weren't taken seriously until the 1960s, with the burgeoning environmental movement. The first and most obvious concern was "thermal pollution," given that it was well known that most power plants discharged huge volumes of heated water in a "thermal plume" that had the potential to block or alter the movements of fish life. But the absence of obvious plume effects allowed the emphasis to shift quickly to a less-apparent threat—two ways in which fish could be killed, called impingement and entrainment. *Impingement* is the impinging of fish by water pressure against the screens designed to keep them out of the bowels of the plant. *Entrainment* is the carrying of fish too small to be impinged right into those bowels. In the first case a fish may be held in place or squashed till it dies; in the second a fish egg, larvae, or small juvenile may pass back to the river, but it will be in "cooked" form.

What first became apparent in the challenge of routing a portion of the Hudson through the guts of a nuclear plant was a deadly thronging of adult striped bass attracted to Indian Point's hot-water discharge. In his classic *The Hudson River: A Natural and Unnatural History,* Robert H. Boyle chronicled the battles fought by grassroots fishermen against the hubris of the utility, Consolidated Edison, and of government agencies, as they sought to expose a design flaw at the Indian Point plant's hot-water discharge configuration, one so extreme that each day thousands of adult striped bass were killed and hauled in trucks to a secret dump, which when exposed grew into a public relations disaster. The existence of the dump only became known when great numbers of crows gathered there. Various unsanctioned photos of the carnage circulated; in one, dead striped bass were stacked a dozen feet high, in another, "The fish seen here were supposed to be about one or two days' accumulation. They were piled to a depth of several feet. They covered an area encompassing more than a city lot."

The core of the early battles over electric generation in the Hudson Valley involved a proposed pump storage plant that would have blasted off the top of Storm King Mountain in the scenic Hudson Highlands, and the environmental impacts of the Indian Point nuclear plant. The politics were rife, but the scientific concerns boiled down to how many fish would these plants kill, and how would these mortalities affect their populations? The first question is answerable within reasonable error, but some results seem shockingly high; the second raised questions that remain controversial today.

How many fish are killed in the operations of a large once through–cycle power plant? Better get a big abacus. For the Indian Point plant in 1987, just after mitigative reduced-flow measures were used, the estimated kill of all eggs, larvae, and other life stages of bay anchovy nonetheless was 227 million; full flow might have killed up to 460 million. That year, of the next top six species entrained, four were bona fide anadromous fishes—striped bass, shad, river herring, and rainbow smelt. That same year impingement in summer alone was estimated to have killed more than 14,000 larger striped bass, along with 24,000 river herring and 2,300 shad. Altogether, the working assumption now is that Indian Point entrains 1.2 billion fish per year and impinges a little over a million. But Indian Point is just one plant, albeit a big one. Farther down the estuary, near Staten Island, the Arthur Kill plant is estimated to entrain 1.5 billion fish and impinge almost 4.5 million more; two power stations in the East River are in the same ballpark, and there are others that also take a major toll. It's expensive to make these estimates, and after a while, the results seem redundant; it comes down to orders of magnitudes of dead fish that utilities operators and regulators agree to use in their negotiations. Nonetheless, the vagaries surrounding them keep a bevy of consulting and staff biologists and lawyers employed.

The numbers of fish that die in the course of this electric generation in this one estuary is simply staggering. The question is, how much does this so-called industrial fishing matter? It depends on whom you ask. Early on, if you asked federal scientists what percentage of the Hudson's striped bass population would be lost as a result of operating Indian Point Unit 2, they would have said 50 percent. But if you asked the utilities' consultants, they would have said less than 5 percent. The discrepancy had much to do with the model used of the river's circulation, but the real core issue throughout these conflicts has been the remarkably thorny question of biological compensation.

Compensation is a biological notion that makes perfect sense in theory but is nearly impossible to quantify. Proponents of compensation suggest that mortality is density-dependent, meaning that deaths of early life stages open up more "room" ecologically for the survivors. Therefore, if power plants kill large numbers of eggs and larvae, their losses are somewhat compensated for by the higher survival of the remaining eggs and larvae that won't have to compete with them for resources. To look at it another way, they are arguing that a fish is not a fish is not a fish, but that an early life stage of a fish killed by a power plant is only, in effect, a fraction of a fish. The problem is, are they 0.9, 0.8, 0.5, or some other fraction?

Decades of research and argument on these issues in the Hudson has led to some concurrence that the combined effect of all the power plants on the Hudson reduces the cohort sizes of striped bass by about 20 percent. Although this is less than the effects of fisheries harvests, it is clearly significant, as are the losses of many other species. We pay more than the cost of our monthly utility bill for our electricity.

Chapter 10

Precautionary Principle vs. Principally Not Cautious

It's not fish ye're buyin', it's men's lives.

—*Sir Walter Scott*

CAPE COD HERRING FISHING, NORTH HARWICH, MASS. jumps out at me as a keeper as I'm sifting through vintage postcards at an antiques shop; indeed, it's a concise and remarkably encompassing rendition of both the high promise and singular vulnerabilities of a fishery for an anadromous fish illustrated in just a single image. A little research on dating postcards suggests it's from the 1930s. Look closely at the scene and you see a quiet creek, the Herring River, leading to a low barrier, easily passable by the river herring. Behind this barrier is a round holding pool with a sloping beach on its business side where a net can be hauled. Above the pool is a higher dam with a large pond where the river herring can pass to spawn. Alongside the holding pool is the end of a long shed that protects dozens of wooden barrels in which to transport the catch by horse and wagon. Later investigation informs me that the fish from this river were used for food, bait, and fertilizer, with some being pickled and the roe prized, and that even the scales were used to make buttons for women's dresses. The herring ganged in the holding pool until each individual either was netted and sent to market or was allowed to pass into the upper pond to spawn. The little holding pool in its best years produced 3,700 barrels of alewives—not a small number of fish when you learn that alewives pack around 400 to the barrel.

Could you ask for any more cooperation by an edible wild creature? The fishermen know based on past experience that the run will begin, for

Cape Cod Herring Fishing. North Harwich, Mass.

"Cape Cod Herring Fishing, North Harwich, Mass."
Postcard

example, somewhere around March 20, so they check the creek occasionally before that, and when the fish show in reasonable numbers, they are ready to intercept them. The river herring, who days before would have been impossible to locate in the vastness of the sea, are now densely concentrated in a convenient and easy-to-net location. But the fishermen also know roughly when the run will taper off, so they can plan other activities such as farming around this knowledge.

With all this control over the fate of these fish, this is where wisdom is required. If the fishermen are greedy and take too many herring, thereby not allowing enough to pass into the pond to spawn, then when their progeny return to procreate in three or four years, the run sizes will be low and those past profits will come at the expense of the later ones. If the fishermen are wise, allowing a significant portion of each week for the spawning fish to pass freely through the holding pool into the pond, the fishery will be sustainable year after year. Early on, the ratio of days fished versus days where fish were allowed to pass was probably set through trial and error and a "feel" for the run. Today we might use sophisticated computer modeling to answer this question, but regardless, the principle is the same: Each year's fishery must leave enough of the run to prosper as an investment in the future of the fishery. Seems simple, doesn't it?

Commercial fisheries for the Atlantic's wealth of anadromous fishes once took many forms, most now either gone or severely reduced in scale. In Canada's Bay of Fundy, fishermen drove long rows of brush into the miles-long mudflats to guide shad, striped bass, salmon, and sturgeon at high water into depressions where they were stranded at low tide. In Maine's Kennebec River prodigious numbers of smelt were harvested in bag nets to be sold in Boston and New York City. In New Hampshire's Exeter River, weirs were used to guide alewives straight into waiting nets. In Massachusetts "pinhookers" (i.e., commercial fishermen who fished with rod and reel) landed massive numbers of hefty striped bass for the market. In Rhode Island the vertical wooden stakes of pound nets steered vast numbers of "schoolie" stripers swimming along the coast into pens that held them long enough to be dip-netted out into skiffs. In Connecticut baymen made holes in the winter ice and "raked" eels out of the mud. In New York, across from Manhattan, shad were caught "in the shadow of skyscrapers" every mile or two on up in the Hudson River to near the middle of the state; shad also approached New York's center through the alternative route of the Susquehanna River, where along the way they helped feed Pennsylvanians. In New Jersey baymen of Barnegat Bay speared eels and trapped them in wicker baskets. Off Delaware, trawl nets made large "intercept" catches of American shad traveling to their natal rivers to spawn. In Maryland watermen caught endless quantities of small "rock" in their nursery of Chesapeake Bay before they became large enough to migrate away. Shad drove headfirst into staked gill nets in the Virginia waters of the Potomac. In North Carolina's Roanoke River an ingenious device called a fish slide was used to strand downstream passing fish such as spawned-out shad. Anchored gill nets in South Carolina's Winyah Bay took large numbers of Atlantic sturgeon for their flesh and caviar. In Georgia fyke nets landed blueback herring and striped bass moving upstream. And in Florida's St. Johns River, haul seines were used to beach mixed bags of American and hickory shad.

Controlling fishing is not simple, however, because it's not fish that are actually managed—it's people. And the people and their interests are divided into many conflicting fiefdoms: individual fishermen (a fiercely independent lot); fishermen who fish a particular gear type or species; groups of fishermen in a geographic region; local, state, and federal government concerns; and the competition between commercial and recreational fishermen.

This confounding societal patchwork is further complicated by the migratory behavior of the anadromous fishes themselves. This issue has played out mainly at the state level, where for a long time the management realm along the Eastern Seaboard was divided between "producer" and "consumer" states. We've come a long way since Melville stated in *Moby-Dick* (in reference to whales, which were considered fish at that time) that "A Loose-Fish is fair game for anybody who can soonest catch it." Producer states were those that had spawning waters for a given species, thus achieving a karmic balance for also harvesting those species in considerable numbers, along with a greater sense of responsibility to their sustainability. Consumer states were those that had access to the adults along the inner three miles of waters of their coasts but that lacked riverine spawning areas (e.g., New Jersey); they were often at odds with producer states. These battles gave rise to something that sounds like it might be a federal agency, but is not: the Atlantic States Marine Fisheries Commission, or ASMFC.

The ASMFC is actually a compact of the fifteen Atlantic Coast states formed in 1942 under the notion that "fish do not adhere to political boundaries." Its mission statement says that it exists "[t]o promote the better utilization of the fisheries—marine, shell, and anadromous—of the Atlantic seaboard by the development of a joint program for the promotion and protection of such fisheries, and by the prevention of physical waste of the fisheries from any cause." As for Congress, no state can exert undue influence because the Commission is made up of three members from each state, with divergent interests, including the director of the state's fisheries management agency, a member of the state legislature, and an individual appointed by the governor, all supported by technical and advisory committees and numerous staffers. The ASMFC now manages twenty-three species, and the consensus seems to be that while the agency is not perfect, it is far superior to the free-for-all that would exist in its absence. The ASMFC has also worked well enough to inspire creation in 1949 of the Gulf States Marine Fisheries Commission. Its five states include Florida, which gets to be on both commissions.

For a brief time the world went caviar-crazy. Perhaps the most heedless fishery that ever occurred for any anadromous fish of the Eastern Seaboard was for Atlantic sturgeon in the Delaware River in the late 1800s, the height of the international caviar craze. Although the Delaware was not the largest river by discharge along the coast, it was nonetheless a sturgeon Shangri-la. With a sprawling estuarine bay and a long, deep riverbed suited to the sturgeon, and with it lying right in the middle of the species' marine range, it supported a disproportionately enormous lode of the whiskered behemoths. Fishing on the Delaware's stock was light until the 1870s, when technology transfers from Europe for fishing and processing caviar provided both the means and the incentive. Peak harvest of Atlantic sturgeon in the United States occurred in 1888, with 7 million pounds landed; despite the fish being found in American rivers from Maine to Florida, almost 6 million pounds of that came out of the Delaware alone.

It was a rape, the piscine equivalent of clear-cutting an old-growth forest. By 1897 the fishery engaged nearly a thousand fishermen among a number of fishing camps, and the town of Caviar, New Jersey, was born, soon becoming the leading provider of black gold in the world. The camps operated similarly to the shad batteries of the Susquehanna; men lived on scows or houseboats where they processed eggs from fish caught close by. Because of the value of caviar, the larger-size females were the targets, and they were nine out of every ten caught with the purposefully fished wide-meshed gill nets. Sturgeon meat was smoked and marketed domestically, mostly in New York City. Caviar was also shipped to New York, but most was then exported by the firm Dieckmann & Hansen to Germany. Fifteen train cars of sturgeon products left Caviar each day for New York. In Europe a portion of the Delaware River caviar was sold as premium "Astrakhan Caviar" from Russia. Some of the ersatz Russian caviar was then re-imported to the United States and sold at the then-astronomical price of six cents per ounce.

Sturgeon are remarkably slow to propagate, so as the harvests mounted there was only modest replacement with new individuals. The army of fishermen instead was exploiting decades of accumulated biomass; their total weight at the beginning of the intense phase of fishing was estimated at 48 million pounds. And the fish were huge; mean weights for Pennsylvania landings were 168 pounds; for Delaware, 185 pounds; and for New Jersey, 278 pounds. Yet, because females don't return to spawn for two to six years after their previous spawning, the fishery persisted for about a decade as the population was whittled down by about 10 percent per year. In 1890 fishermen landed an average of sixty sturgeon in a net haul; in 1896 this dropped to twenty-seven; and by 1898, only eight. As is common, though, with the naked market forces of supply and demand in fisheries, a 135-pound keg of caviar was worth $9 to $12 to a fisherman in 1885, $40 in 1894, and $105 in 1899. By 1901 the fishery had crashed, with Delaware and New Jersey landings only 6 percent of their peak levels in 1889. Atlantic sturgeon would remain almost untraceable in the Delaware—the river where they once flourished as if in a natural hatchery—for more than a century. And the town of Caviar was renamed Bayside.

The worldwide history of preventing overfishing on important fish is a sad one, and the record for restoring stocks that crash is also nothing to be proud of. For Atlantic diadromous fishes there is one exception: migratory striped bass. Striped bass of two Mid-Atlantic estuaries, the Chesapeake and the

Hudson, had long supplied the many coastal recreational and commercial fisheries from Cape Hatteras to Maine. The societal importance of these fisheries cannot be overstated: Commercial fisheries of many varieties, from hook-and-line through gill nets and haul seines, took enormous numbers from the many states where commercial fishing was allowed. Charter-boat captains made livings taking people out to land their first "striper," and armies of dedicated anglers rode their own boats out to ocean reefs, rocky bays, sandbars, estuaries, and brackish rivers in search of them, often in sight of countless anglers who would throng to the shore day and night when word went out that "the stripers were in."

Though not truly cyclic, striped bass numbers of the Atlantic migratory populations are highly variable. Even as early as the mid-1700s, concerns were expressed about scarcities being related to "very great numbers having been imprudently, or rather wantonly taken in one season." But so-called dominant year classes would appear reliably every few years, like a giant hemoglobin injection into the vitality of the stock. An exceptionally large year class was spawned in Chesapeake Bay in 1970, and these fish helped to drive commercial landings to a historical high of almost 15 million pounds.

The 1970 year class and the others of that time were fished relentlessly, and unlike in other decades another dominant year class was not forthcoming. As a budding striper angler on the shores of the Bronx in those days, I found that the fishing was easy and the keeping could be gluttonous; you could harvest as many as you wanted to, and the only regulation was that a striped bass had to exceed sixteen inches, which occurs at little more than two years of life past their hatching. And size limits were even more liberal farther south, only ten or twelve inches. As the 1970 class grew, though, more and more large stripers were caught by the close of the decade, exciting sportfishermen who sought—and some who caught—the fish of a lifetime. But by 1983 commercial landings had declined to a little more than 2 million pounds, and alarms were being sounded. Clearly a crash was occurring, one that would ripple through the economies of many coastal states, where commercial and recreational striped bass fishing was a way of life. Things by then were so bleak that two Maryland charter-boat captains even filed a petition to list striped bass under the Endangered Species

Act, a largely symbolic gesture that, nonetheless, shook up the management realm.

The ASMFC stepped into the crisis with the Interstate Fisheries Management Plan for Striped Bass, and so did Congress with the Emergency Striped Bass Study, funds leveraged by Rhode Island senator John Chafee. First came substantially longer length limits and spawning-area fishery closures during spawning season between 1981 and 1984; when these measures didn't work, amendments to the plan indicating further restrictions were added. One problem was that with the diminished stock, not enough eggs were being spawned even on those years when the ecology was favorable to their survival. Most critical to remedying the dearth of eggs was Amendment 3, the brainchild of John Boreman, a talented fish population modeler with the National Marine Fisheries Service who'd been handed the challenge of a lifetime to restore these iconic fisheries. Boreman's strategy featured the 1982 Chesapeake Bay year class. Unlike the 1979, 1980, and 1981 abysmally small cohorts, the 1982 class at least was near the long-term average. Boreman's insight was that if almost all fishing pressure was relieved on only a modest year class, that year class by the time of spawning would be equivalent to a dominant year class.

Last-ditch draconian measures were ordered. Along a coastline where striped bass had been rampantly harvested for centuries—including the Chesapeake, where tradition allowed the keeping of untold numbers of foot-long "pan rock" (little stripers that would fit into a frying pan)—states were told they should protect 95 percent of the females of the 1982 year class until 95 percent had the opportunity to spawn at least once. The states responded with their own combinations of restrictions, including length limits as high as thirty-eight inches, and total moratoria on striped bass harvests. The plan made sense on paper, but would it work in the water?

Vorsorgeprinzip may mean little to you unless you speak German, but if you promulgate a fishery, this socio-legal 1930s concept may be the new philosophy of how you do business, translated today as the "precautionary principle." During the 1990s there was a sea change in the broader out-lines of fishery management, a switch from external forces needing to dem-onstrate that a fishery was harmful, to the fishery itself needing to show that it is not—what many see as a righting of international ethical order. This "trial without error" seeks to accommodate both insufficient scientific understanding and the need to avoid significant and, especially, irreversible injury. Critics of the principle claim that it deters progress and develop-ment, is excessively risk-aversive, and is unscientific. But the precaution-ary approach is beginning to take hold for high-seas fisheries at the level of the United Nations and the Food and Agriculture Organization, and not a moment too soon. Despite the evolution of a more-consultative philosophy, however, it has not yet been fully adopted for diadromous fisheries on the US Atlantic coast. What's left of our eel fisheries is a case in point.

Globalization has brought an irony to American eel fishing. In a nation where eels helped to sustain our colonial ancestors, most citizens today would not think to eat an eel, and because of this, there is little targeted fish-ing for adult eels. But those who still hold the right—and even those who don't, but are nonetheless tempted—can't catch enough baby eels because of the extraordinary prices on their heads. There was a time when young eels entered the rivers of the Atlantic Seaboard unmolested, where to see them might not elicit anything beyond awe from both their numbers and their sheer determination to fight and climb their way past obstacles upriver. One observer wrote of elvers ascending streams: "In the course of the sum-mer young individuals ascend rivers in incredible numbers, overcoming all obstacles, ascending vertical walls and floodgates entering every large and swollen tributary, and making their way even over *terra firma* to waters shut off from all communication with rivers." A newspaper account tells of a Mr. Wallace who was informed by his wife, who had gone to the Big Bushkill in Pennsylvania for a pail of water, that there was a mass of eels ascending the creek. Mr. Wallace went to the creek and for a while "watched a procession such as he had never seen before." The four-inch-long eels formed a dense column up to three feet wide that was rapidly making its way upstream. Mr. Wallace left and returned an hour afterward and found the line still going. Remember that this was just one of thousands of spurs glass eels could turn off to after traveling for months with the Gulf Stream and then deciding to turn toward the coast.

Today these massive incursions by eels into American waters are no more. It's as bad in Europe. By the mid-1980s the number of glass eels entering rivers there declined to 10 percent, and by the end of the twentieth century, to 1 percent of historical numbers. The crash occurred all over Europe, with no single, obvious cause. In the United States eel abundance was fairly steady until the 1970s, but has declined at a rate of 12.5 percent annually since. Try applying that kind of loss to your IRA for forty years and see where you stand. In fact, the remaining American freshwater population is at less than 1 percent of historical levels. But eels are still in huge demand in Japan and other Asian countries. Want to become incredibly rich? Figure out how to spawn eels in the lab in commercial quantities. No one has succeeded. So as glass eels become scarcer and scarcer in the United States these days, Asian demand only increases for young eels that can be grown to market size in captivity.

Further review by ASMFC should, by all rights, shut them down, but as of 2013, two states still allow fishing for elvers: Maine and South Carolina. Not because it was clearly sustainable, but because they had long-standing fisheries and so were "grandfathered." From the early 1980s to 1992, the market for wild Maine elvers was small. But since 1993 the worldwide harvest of wild eels has been unable to satisfy the market, raising prices and attracting more fishermen to Maine's tidal rivers. Supply and demand, illustrated. It takes about 1,500 baby eels to round out a pound. In the early 2000s the price of elvers was about $25 per pound; recently, $250 to $300 per pound was considered a high price, but in 2011 it shot to $950 per pound. On a per-ounce basis, this is about double the 2011 value of pure silver. But it didn't plateau there; the 2011 tsunami in Japan and a crackdown on eel exports in Europe caused the price in spring 2012 to open at more than double, $2,200 per pound, while the value of silver plunged. Eels were by then worth five times the value of silver.

The legal eel fisheries have a predetermined season. The *New York Times* reported that the 2012 fishery in Maine opened to a strong run; one exporter said, "The first two days of the season were extremely amazing," and that "people were making $30,000, $40,000 a night." A friend described it as a gold rush. At these numbers the temptation in Maine for the unlicensed many to fish illegally among the licensed few (400 legally licensed) is great, and Governor LePage signed an emergency measure to raise the penalty from $500 to $2,000, which even at that level probably isn't dissuasive; one buyer claimed that "poachers were coming out of the woodwork."

Although many elver fishermen have private spots, others cluster on good river reaches. To stake out the best positions, some sleep on-site in their trucks for days before the season opens. Someone who put in three days camping on a river said, "I had to actually let another guy take the spot; otherwise, he was going to fight me for it. And I wasn't going to go to jail, knowing what kind of season was ahead of me." Big money for little fish. An exporter said some dealers are "walking around with security because of all the cash they carry from buying elvers in the middle of the night." Once, seemingly endless numbers of economical mature eels helped to feed new immigrants to America from other countries; today the dwindling young eels that attempt to pass into certain US waters are hijacked to be sent as a luxury item to another country.

Chapter 11

Concrete Crimes against Rivers

If you dam a river it stagnates. Running water is beautiful water. So be a channel.

—*English proverb*

In 2002 John McPhee published his paean to shad, *The Founding Fish.* Thirty-one years earlier he put out another book, *Encounters with the Archdruid,* in which the final third was devoted to the shad's number-one enemy: dams. In discussing arch-environmentalist (to some, with druid-like tendencies) David Brower, McPhee eloquently put dams in a conservationist's perspective among a litany of environmental insults: "The outermost circle of the Devil's world seems to be a moat filled mainly with DDT. Next to it is a moat of burning gasoline. Within that is a ring of pinheads each covered with a million people—and so on past phalanxed bulldozers and bicuspid chain saws into the absolute epicenter of Hell on earth, where stands a dam. Conservationists who can hold themselves in reasonable check before new oil spills and fresh megalopolises mysteriously go insane at even the thought of a dam. . . . [P]ossibly the reaction to dams is so violent because rivers are the ultimate metaphors of existence, and dams destroy rivers."

If dams destroy rivers, why do we have so many of them? Well, flood control, mechanical energy, electrical energy, water for irrigation, drinking water, recreational opportunities. These valued and often-essential services of dams seemingly rank them among mankind's greatest inventions. Maybe even give da Vinci a little recognition if you are a dam lover; he planned one that would have extended all the way from the sea to Florence to sustain water in a canal. But the truth for dams is mixed.

Drive along the Eastern Seaboard and you will see dams and more dams. They come in many forms: arching and double-arching, buttress, earth and rockfill, along with gravity dams that rely on their weight for support, among others. Many roads follow river courses, and there are more roads near population centers, and population centers support dams, and vice versa. Exploring a scenic two-lane highway in northwest Connecticut, I spot Northfield Brook Dam, a landscape feature that looks positively Hooverian for this part of the world at more than 800 feet long and higher than 100 feet, colossal at its base because it's made of rolled earth and covered with loose rock, like talus on a mountainside. But when I park at the access lot at the top of the dam to look at the lake behind it, there almost is none. Looking down into the great artificial hollow, I see nothing more than a modest pond far below, fed by a brook so small that a single railroad tie could dam it most of the time. Why impose so much dam on such little water?

In 1955 this part of New England experienced two hurricanes in one week. On August 16 Hurricane Connie dropped six inches of rain, followed five days later by Hurricane Diane, which dumped another fourteen inches—so much rainfall in such little time that even brooks became rivers and left parts of Connecticut devastated. In response, ten years later the Army Corps of Engineers finished the Northfield Brook Dam project, conducted at a time when there was little questioning of decisions to engineer the environment on whatever scale the agency saw fit. The vacancy behind the dam is storage capacity if and when another event like this ever happens. Doubtless some area residents with long memories enjoy the feeling of safety the dam provides and the little beach where they can swim, while I suspect others view it as a massive overreaction, an out-of-scale blight on a bucolic valley.

Although electric hydropower is more closely associated with giant western dams like Hoover, many rivers that drain to the Atlantic also generate watts, but on a smaller scale. Visit the guts of a major hydropower installation, such as the Conowingo Dam on the Susquehanna, and you'll hear the anxious roar and feel the trembling as biblical quantities of water are forced through giant machinery to work for the people. The energy of water flowing downhill spins rotors in turbines that zip electrons into our power grid. Clean and sustainable energy, as it's often billed? Perhaps. But only if you view it in the same way politicians too often view gambling casinos. Yes, gambling adds a new revenue stream, but what of the hidden expenses—societal costs such as addiction, bankruptcies, foreclosures, and broken homes? Hydropower dams hum quietly in the surroundings, little noticed, but wreaking unseen biological havoc on an ongoing basis.

Before electric hydropower there was mechanical hydropower. Travel almost anywhere in New England and the Middle Atlantic States and you will see abandoned milldams liberally dotted across the countryside, the most recalcitrant pieces of our industrial legacy. The mills are mostly long gone, along with the accompanying infrastructure and the workers and their paychecks, but many of the river-stopping earth, wood, and cement barriers at the heart of these enterprises remain. Some still back up rivers and form pools with parks or private waterfront backyards that have their local constituencies for various stillwater activities. All milldams, of course, have histories, and some are appreciated for their historical value. But many are useless, some heartbreakingly so from an ecological point of view, such as dams in freestone rivers that over time have accumulated enough rock rubble on their upstream sides to reach the top of the dam, with occasional stones tumbling over the divide.

Water flow is a "master variable" of rivers and streams, and dams do nothing but disrupt natural flowage and the structure and function of eco-systems. Dams also serve many purposes and have varied constituencies. Pile the pros and cons on a societal scale, and some dams come out largely justified in their continued existence; for some, their value is debatable; and some remain only through inertia. It takes considerable effort and resources between selling the notion and gaining approval for it, to then actually remove the great mass of a dam.

A dam is the most profound affront to the ecological health of a river. This bears repeating: *A dam is the most profound affront to the ecological health of a river.*

Among their most notable consequences: Dams stop rivers from "running silver" with anadromous fishes. The effect often is not complete; if there is enough river left between the dam and the sea, or if fish-passage devices are at least somewhat effective, or if hatcheries dump enough artificially reared young fish into the river, runs may persist. But often these runs are nothing more than sad relictual echoes of the natural abundances of the earlier pristine runs. Though the blocking effects are well known, the reactions by the fish that first encounter them are usually just surmised. But rare observations were made after the Veazie Dam on the Penobscot was completed in the 1830s. Two Maine fisheries commissioners later recounted: "When the fish came in the spring they found an impassable barrier across their way; they gathered in multitudes below the dam and strove in vain to surmount it; many returned down the river; and after the usual time for spawning of shad was past they were taken in weirs in the town of Bucksport, loaded with ripe spawn they could no longer contain; a phenomenon which Mr. John C. Homer who has fished with weirs at that point for forty-three years had never observed at any other time. These were doubtless shad whose natural spawning grounds lay far up the river, and who had after long contention given up the attempt to pass the Veazie Dam. A great many shad and alewives lingered about the dam and died there, until the air was loaded with the stench."

Dams also make it more difficult for any outmigrating juveniles produced to move back downstream. Not only must the young wash over the top of the dam, struggle to locate and take a ride in some sort of special bypass chute, or pass through hydro turbines, but the downstream sides of dams often aggregate predators that pick them off as they migrate.

And although the slowing and stoppage of migrating diadromous fishes is most noteworthy, dams prevent the assemblages of residential fish and crustaceans and other aquatic species from making their normal movements within a river. Some of these journeys are associated with reaching spawning grounds, others with seeking the appropriate temperatures, flows, or refuges from predators, but all are important. Dams critically impede populations of ecologically important, filter-feeding native freshwater "mussels," which actually are clams that live buried in river sediments. Being immobile and lacking a means to get their young life stages up rivers, some have even evolved elaborate fleshy appendages that look like prey that they wiggle to attract feeding fish. When the predator draws near, the mussel blasts them in their head with a spray of parasitic larvae. Two now sharply reduced mussel species are specialized to rely, one each, on alewives and eels to carry their larvae upstream. So limiting is this access to the proper host species that eels are being trucked past the Conowingo Dam on the Susquehanna and released upriver to help save the eastern elliptio mussel.

Also, by fragmenting populations, they reduce the long-term prospects for survival of larger, less-numerous species. Each population segment between dams becomes the equivalent of an island population, and the smaller abundances of these isolated fragments are less robust in coping with stresses such as overfishing, predation, and disease. With dams the natural free-flowing continuum becomes a series of chopped-up segments where the whole is a great deal less than the sum of its parts.

Besides being a physical barrier to migratory fishes, dams also change the greater geophysical regimes of rivers. Reservoirs heat up faster, raising river temperatures. But some larger dams release water from lower levels of reservoirs, waters that because of stratification are often much colder than they should be—the river discontinuum concept illustrated. Dams also change the hydrological signal to those fish that achieve fish passage; after tracking the flow to reach the dam, they may find themselves in a vast water body with no or little discernible current, causing them to waste time and energy seeking the reservoir's inflow. Additionally, the diminished river currents caused by dams turn them into sediment traps: Sand, mud, and clay drop to the bottom of quiescent reservoirs instead of traveling downriver. On the Susquehanna suspended sediments upriver at Harrisburg, Pennsylvania, exceed sediment discharge at Conowingo, Maryland, near the river mouth, probably due to sediment deposition in the lower river and behind three hydroelectric dams. These trapped sediments and the incorporated nutrients they carry don't make it to a river's estuary, where they would normally play an important role in the food chain. Depending on circumstances, this can be desirable (where there is nutrient over-enrichment) but finite, or undesirable (in places that become nutrient-starved).

Dams even directly change the chemistry of rivers. Hard-shelled algae called diatoms live in reservoirs, their shells composed largely of silica. As the diatoms die and sink to the bottom, they take much of it in the system with them, starving estuarine and inshore coastal waters of the silica needed for their diatoms, an important part of the base of the marine food chain. The resultant phytoplankton community is different than the normal one that would exist if the river ran freely. The scale for which this can occur is huge; silica held behind the "Iron Gates" Dam on the Danube has changed the character of plankton production a thousand miles downriver in the Black Sea. The fundamental realms of the river environment—biology, physics, chemistry—are all altered in myriad ways that are detrimental or fatal to the organisms and communities that evolved in the absence of dams.

A river with the profound affront of even one dam on it is highly compromised in comparison with a free-flowing river. So how many dams are there? The number of dams already built is staggering. By the year 2000 the US National Inventory of Dams recorded more than 80,000 dams that are six feet or higher, but the actual sum of dams inventoried by the states is more than 99,000. And state officials are constantly discovering previously uncounted dams during routine inspection trips. Some believe that several million dams of all sizes may dot the nation. Dams are so common in the United States that only forty-two still-high-quality but undammed rivers of at least 125 miles in length still persist. And it didn't take long to dot Atlantic rivers with dams; a recent compilation of US manufacturing census data showed more than 65,000 water-powered mills by 1840. Even today the Hudson River watershed alone contains at least 797 dams.

One of the most powerful and easily understood metrics for the effects of dams on diadromous fishes is the sum of river miles and habitat lost to spawning runs because of them. Karin Limburg estimated that American shad have yielded about 2,500 miles of rivers due to dams, about 35 percent of their total. Maine saw construction of its first sawmill dam in 1634. As of 1829, an estimated 1,686 manufacturing establishments depended on water power in the state. By 1870 state fish commissioners concluded that dam construction was the principal cause of migratory fish extinction from Maine's waterways. Twenty years later, they stated that only 10 percent of the original habitat in Maine was available for spawning because of damming.

Historical ecologist Carolyn Hall and her colleagues examined the history of dam building and its consequences for alewives in Maine. Dams peppered the state. Some systems did not have many dams, but only in the relative sense. Short, low-relief coastal rivers such as the York, Fore, and Pemaquid had a dozen or fewer, but the great Kennebec system held 226, of which 128 still stand. And the even-larger Penobscot had 283, with 116 extant. The consequences for alewives and other diadromous fishes were dire. Historically, alewives migrated 200 miles up the Penobscot and 120 miles up the Kennebec. But after the lowermost dams were built, the percentage of stream habitat still available to alewives was less than 20 percent for the Penobscot, and only 7.3 percent for the Kennebec. Worse yet, alewives normally spawn in lakes at the heads of rivers. When Hall and her colleagues looked at the reduction in combined river and lake habitats in Maine, it had been reduced by 97 percent by 1900. How did habitat loss translate to fish numbers? Appallingly so, of course. For nine obstructed watersheds, alewife numbers were reduced to between 0 and 16 percent of virgin estimates, a cumulative lost-fisheries production between 1750 and 1900 of 11 billion fish!

The postindustrial river has far fewer eels swimming within it than in prehistoric or colonial times, and dams are no small part of the reason. As much as 84 percent of the historic habitat for eels in East Coast watersheds and in the Lake Ontario Basin is no longer directly accessible because of dams. Ever since the Lake Ontario watershed deglaciated, enormous runs of eels passed up and down the pathway to it from the ocean, the St. Lawrence River. It is conceivable that these runs even expanded when eels gained access to Lake Erie and the three other Great Lakes with construction of the Welland Canal in 1829. Netters had long harvested immense numbers of big silver eels moving downstream on the St. Lawrence—eels that had grown and fattened in the big lakes themselves, and all the way to the heads of their sprawling web of tributaries that infiltrate much of the northern states and the Canadian Shield.

In 1958 the ninety-foot-high Moses-Saunders Power Dam came online on the St. Lawrence, flooding ten communities near Cornwall, Ontario, that came to be known as the Lost Villages. It also was a losing proposition for the American eel, followed by twenty-six years of construction of another dam on the St. Lawrence, at Beauharnois, Quebec. Perhaps one of the few pluses to placing obstacles such as dams in the path of migrating fish is that they can provide a means to count those passing through. But because of the detrimental effects of dams, these counts also rarely offer good news. Since the early 1980s, biologists have counted young eels moving up an "eel ladder" that zigzags its way over the Moses-Saunders Power Dam. In the mid-1980s, between 25,000 and 30,000 pencil-size eels a day slithered up the ladder and over the dam as they moved upriver toward the Great Lakes. "It was just seething with eels," said John Casselman, an eel expert with the Ontario Ministry of Natural Resources. The population of young eels coming up the ladder plummeted from nearly a million in the 1980s, to about 100,000 in the early 1990s, to less than 10,000 in the late 1990s, to near zero in 2000. Today, peak numbers are only 20 or 30 a day, Casselman said, and those eels are not young eels moving upstream, but larger ones simply moving back and forth on the river. Lake Ontario was once dominated by eels, with female eels perhaps constituting one-half the flesh of its inshore fishes; today, nary an eel is to be seen.

Upstream of the dam, Lake Ontario alone was home to 5 to 10 million eels as recently as two decades ago. That number has declined to several tens of thousands as old eels migrate out and are not replaced by young eels. "It is like our passenger pigeon," Casselman said. In Oneida Lake, in the Lake Ontario watershed, in the early 1900s, one hundred tons of silver eels were taken annually; after dams and canals were built on the Oswego and Oneida Rivers, only two eels were caught in research sampling in Oneida Lake over a twenty-five-year period.

Even if the fish ladders on the St. Lawrence River are effective at passing eels upriver, the dams they serve are ruthless in their effects as the fish move downriver again, there, and at many other hydroelectric stations down the coast. The turbines spinning at both St. Lawrence dams together kill about 40 percent of the large silver eels aiming for the Sargasso, and more may be sufficiently wounded so as to be unable to complete their great hemispheric journey.

The crash of eels at the well-monitored St. Lawrence with its eel ladders shows a dismal pattern; we can only guess at the effects of dams that lack any provisions for passage. Although eels have mysterious means of passing many dams that would seem difficult if not impossible for them, each nonetheless acts at best as a filter. My colleague Robert Schmidt of Bard College at Simon's Rock measured eel densities behind a series of dams on a small tributary to the Hudson River. Though eels were found all the way upriver, he proposed as a rough rule of thumb that each dam reduced eel numbers by about 90 percent. But this applies to dams that are at least somewhat apprehensible to eels. Before the giant Conowingo Dam blocked almost the entire Susquehanna, annual eel harvests were almost 1 million pounds per year. After the dam was built, they were zero.

Construction of a dam is often seen as a solution to flood control, but it can simultaneously be viewed as a concrete crime against fundamental principles of watershed management. Two of its effects are clear violations: It interferes with the natural deposition and erosion processes—the very forces that carve and shape a river—and it attempts to override the normal fluctuations in flow, in effect, erasing the consequences of its seasonal and irregular changes. Finally, it hijacks the river channel itself to become the only important conduit, negating the normal role of watershed soil and vegetation storage of floodwaters.

Flood control might be viewed as one of those wonkish subjects populated by mild-mannered engineers who meet quietly and make incremental improvements to what appears to be a mature science. But nowadays residents of New Orleans, Brownie, and George W. Bush might differ. Indeed, flood control equals river control. And to control a river, or at least, to try to, is to wage a monumental battle likely to be least successful when most challenged and, therefore, catastrophic in its consequences. But while manipulating and armoring a river to guard against those ostensibly rare 100-, 500-, or 1,000-year floods, every day their basic ecological functionality is diminished.

Peter Black, a retired Distinguished Professor of Water and Related Land Resources at the Environmental School of Forestry in Syracuse, New York, was invited to a panel discussion at a professional meeting and decided to let decades of frustration and hard-earned wisdom loose on the contentious issue of flood control. An early draft of the resultant paper was titled *Flood Control: Misnomer? Oxymoron? Or Fantasy?*, but a colleague told him this was "too in your face." Black wrote in a summary statement that "civilization's flood control program is not a Good Idea. The evidence is overwhelming." But it's been US federal policy for some time. The preface to the Mississippi Valley Committee Report is topped with the quote: "The ideal river which does not exist in nature would have a uniform flow." This document led to the 1936 Omnibus Flood Control Act, which set federal programs in motion. When the prominent geographer and floodplain expert Gilbert White was asked in 1987 how such an erroneous notion could have become public policy, he blushed and admitted that he had authored that statement. But he also said that it was incomplete and taken out of context, and that he'd actually written that it should *not be* public policy. Nonetheless, the phrase in the report clearly reflected the engineering sentiments of the time, which lived on through decades of ill-conceived river management and not uncommon failures to actually control floods.

Black identified three grand flaws in the flood-control paradigm. The first is the reluctance to accept that floods are natural and useful components to how rivers function. In its upper reaches a river contains much potential energy that normally is dissipated as it courses through and sculpts its channel; thus, a river's size, sinuosity, and floodplain are shaped by its varying flows. Technically, a flow that just reaches its "bank-full" capacity is at flood stage, but it's the waters that rise beyond this that have steered flood-control projects. The natural effects of waters that exceed bank-full are integral to a healthy river corridor: Water, sediments, and organic debris spread out over the land adjacent to the stream that build soil, recharge groundwater, and provide nutrients for plants and animals, all serving to reenergize the river channel, floodplain, and associated wetlands. Timber and limbs washed downstream by high flows may clog the main channel, causing natural diversion of currents to clear out old channels and to establish new ones, braiding the river bottom and creating additional habitat. Floodplains, then, are both a river's safety valve to handle excess flow and also a critical realm of ecological rejuvenation.

Black's second flaw is the lingering influence of the historical viewpoint—the mind-set that as European colonists and their descendants moved westward and "conquered the wilderness," floodplains were attractive and appropriate places to tame and settle in. Floodplains offered rivers for transportation, natural flat corridors for roads, fertile lands for crops, and constant water supplies for domestic and industrial use, for energy generation, and for fighting fire. Floods, until the flood-control programs, were a necessary cost for the enjoyment of all these benefits.

In temperate regions such as the Eastern Seaboard, rivers naturally flood every two to three years, but most of these are minor in scope. Black's third flaw is that more potentially catastrophic floods are assigned centuries-long values, and that there is understandable public confusion about these. A 500-year flood means that a flood of that magnitude or greater can be expected *on average* every half-millennium. But because each one is an independent event, there is no reason why two 500-year floods could not occur a year or two apart, or even in the same year. Moreover, it *seems* as if these supposedly rare events occur pretty darn often. Some of this may be due to the sheer uncertainty in attaching time spans to biblical levels of precipitation, some to a possible ramping up in their actual frequency because of climate change, and some to increasing media coverage of disparate floods that may be a 500-year flood for one region and then for another and then another, all from distinct storms, but making it appear like the skies are suddenly angrier.

Floods are occurring more often despite our sinking more and more federal dollars into control measures. The reasons for this have to do more with the land than with the atmosphere. By continually making floodplains less pervious by building parking lots, playgrounds, and patios, less water is retained by floodplain soils, and more washes directly into rivers, increasing the effects of major storms. It seems that every year we see news footage of distraught homeowners returning to wet and moldy houses as waters recede, saying this is the second or third time this has happened to them, and that it will also be the last—they've had enough. And for those who still want to stay, flood insurance becomes unavailable or unaffordable, an actuarial concession to reality.

Black summed things up when he said, "We don't control floods economically any more than ecologically or hydrologically; floods control us." It's clear we need to wash away our old thinking on floods.

Dams may appear as inert and timeless as the Egyptian pyramids, but they are finite and have expected life spans. But some dams age prematurely, in effect, when they fail conceptually, filling more and more with organic and inorganic matter instead of water. For many dams this happens much faster than the loss of structural integrity of construction materials. The buildup of muck and sand behind Susquehanna River dams has been likened to a time bomb; when their reservoirs become filled with this material, huge amounts of sediments will spill over. In some estuarine systems, the return to normal sediment movements might be seen as a welcome addition to the fertility of the waters, but the Susquehanna flows to the Chesapeake, an estuary where environmental problem number one is *eutrophication*—overfertilization with nutrient runoff from myriad human sources. Expected 50 percent increases in phosphorus and a doubling of turbidity from relaxation of this dam storage effect would counteract minimal but hard-won and expensive improvements in the condition of the East Coast's greatest estuary.

Sediment buildup behind Susquehanna dams brought fear to resource managers in the late 1990s when it appeared some reservoirs might fill within twenty years. There has been something of a reprieve; newer estimates suggest the problem will become acute a little later, around 2025—but even that is less than a decade and a half away. When the four lowermost dams were built between 1904 and 1931, it was thought they could trap 250 million tons of sediments; three have reached capacity, and what's left to sequester some additional sediments is the lowermost dam, the Conowingo.

The Susquehanna's sediment budget is huge and its activity ceaseless. Each second, a whopping ninety-five pounds of sediment flows into the Bay from the Susquehanna. Human activities in the watershed are the culprit. Primeval forests yielded only about one-tenth of a ton of sediment per acre per year, and total inputs to the river were about a million tons per year, compared with about 3 million tons today. Between the pristine beginnings and now, the forests were cleared and plowed, often with furrows that led straight down hillsides, speeding dirt to water. And the earth was mined for coal, with vast amounts of waste flowing to the river, causing sediment transport to reach about 6 million tons annually by 1900.

If good solutions exist for this problem, no one has found one. Backing off on sediment inputs buys time but doesn't remove the deposits already in place. And backing off isn't easily accomplished; modern farming practices strive to reduce erosion, but it is still considered normal to lose about three to four tons of soil per acre per year. One bad solution is obvious: dredging. But the track record for this and the scale necessary to achieve meaningful results suggest this will be far from a panacea, and could be a boondoggle. In 1997 the US Army Corps of Engineers began to dredge a mere 1 percent of the almost 15 million cubic yards of sediment behind a flood-control dam in Pennsylvania. Even though the Corps was able to dispose of the dredged material on-site instead of having to transport it elsewhere, the project still cost $2.3 million. However, the dredging was halted before it was finished when a single storm washed 50,000 cubic yards of sediment back in. Today plans are being made to dredge again, aiming for half the annual loading, and to get rid of it by selling it for construction material. Even if successful, this would only slow down the inevitable. And a "Big Scour," sometimes called a "Katrina-wingo" storm, could still send what scientists call a "catastrophic pulse" of muck from behind the Conowingo Dam into Chesapeake Bay, much like the four years' worth of sediments Hurricane Agnes delivered in 1972.

One analysis of 76,500 mostly medium-size dams in the national database showed that more than 90 percent of them have a median time for their reservoirs to be half-filled with sediment of about ninety years. Smaller dams have reservoirs that are expected to become half-filled with sediment in twenty-five to forty years. Many of these are in the Northeast, which has numerous old dams already filled with sediment. Dams are now being built to accommodate a century of sediment storage, but human activities that loosen soil in the watershed can greatly reduce a reservoir's expected life span.

And dams themselves are growing older and sometimes failing. As dams age, the continuous water pressure and decay of construction materials serve to weaken them, resulting occasionally in reservoirs that need to be partly or fully drained in order to protect people and property downstream. Or reservoirs may suddenly give way because of earthquakes or torrential rains, causing tsunami-like deluges that kill. In 1977 the earthen Kelly Barnes Dam in Georgia failed after seven inches of rain fell, drowning thirty-nine people. In 1996 the Saguenay region of Quebec received eleven inches of rain in a few hours, which caused a network of dams and levees to collapse. Though only ten people died, almost 500 homes were destroyed, and over 16,000 people had to be evacuated. The mother of American dam failures was that which caused the famed Johnstown, Pennsylvania, flood of 1889. Heavy rains took out a thirty-seven-year-old earthen embankment, causing a slug of water to race for almost an hour downstream before slamming the city. Though it traveled as a thirty- to forty-foot-tall flood wave, at one point it reached almost ninety feet. Within ten minutes the city was destroyed, the waters demolishing 1,600 homes and 280 businesses and killing 2,209 people, more than one in every five residents.

In 2005 the American Society of Civil Engineers gave US dams a grade of D, and they're not getting any better. Most dams need major repairs twenty-five to fifty years after they're built, and most US dams are at least a quarter-century old. As they age, their danger increases, not only because of advancing decrepitude, but because of "hazard creep," the tendency of developers to build directly downstream from dams. The irony is that even though Americans now build fewer dams, more people are threatened by dam failures. The Society estimated in 2005 that the number of unsafe dams in the United States exceeded 3,500.

It gets worse. Dam safety officials are so overworked that in most states they don't come anywhere near making all the inspections required by law. According to the Society, the average state dam inspector is responsible for 268 dams; in four states the number exceeds 1,200.

An underlying problem with maintenance for safety of dams in the United States is that the majority, 56 percent, are privately owned. Many private owners can't afford to repair aging dams; some go as far as tying up official repair demands in court to delay or avoid paying for them. Some dam owners walk away from their responsibilities; 12 percent of the dams in the National Inventory have no known title-holder. These stark realities and sober concerns suggest growing opportunities in the future to take dams down for safety reasons and, in the process, as a welcome by-product, to revive the ecology of the rivers they obstruct.

I'm on another drive in the country, this time past East Branch Croton Reservoir, part of the great New York City Reservoir System, a dam-based network originating as far as 150 miles from Gotham that hydrates the city on a daily basis. But I do a double take when I see that the reservoir looks nearly empty; it's down some twenty-five feet, and vast areas of its bottom are showing like a cracked-mud desert. Curious about what's under the surface of this lake I sometimes ice-fish on, I walk down a long valley that contains a small stream that normally empties into the water body at another point far upcurrent.

Remains of temporarily exposed dwellings submerged by East Branch Croton Reservoir, New York
John Waldman

The sprawling Reservoir System, the beginnings of which were placed in service in 1842 because the growing city was running out of potable water, was imposed by eminent domain on the country folk of little hamlets and farms that edged the streams in the suitably contoured valleys that could be dammed. The city's heavy-handedness is still famously resented in the newer Catskills portion of the Reservoir System, where the dammed were chased off their land in the mid-twentieth century. But in the older Croton portion of the system, the landowners were forced to evacuate in the 1880s, and their legacy is mostly forgotten.

As I walk downstream across the shingles of dried sediment, I need to gaze upward to view the fleet of anglers' rowboats ringing the shore, elevated high above my head. But I also start to see echoes of the valley's antediluvian past: first, a small bridge crossing a reborn feeder brook; then a pair of abutments on the main flow, stones neatly piled, of a bridge that would have been large enough for horse-drawn wagons; rock walls crisscrossing the bottom that once edged farm fields; a roadbed that follows the river; a landing on a natural pool where skiffs would have been launched; and the perfectly preserved stone foundations of primitive dwellings. This historical legacy only rarely emerges, during major droughts or when the dam is being repaired. But when it does, it's a strong reminder that at least in human terms, dams are only ephemeral obstructions on the landscape, whereas rivers are forever.

Chapter 12

Climate Change: Latitudes and Attitudes

The planet has a fever.

—*Al Gore*

There continues a loud and politically loaded war over the reality and causes of climate change. More difficult than the comparatively straightforward consideration of the pure physics and chemistry of increasing greenhouse gases, cloud cover, ice melt, and other global driving and feedback forces are the *ecological effects* of climate change on organisms. To test for a climate-change effect on life itself is to be challenged by the reality that plants and animals live in multifarious environments with endless confounding factors. But some analyses have stood out.

Butterfly expert Camille Parmesan was smitten with the Edith's checkerspot, a modest-size lepidopteran with highly variegated wings. As a graduate student in the early 1990s, focusing on their diet, Parmesan noticed more, believing that "You get a feel for the pulse of the species you are working with . . . a kind of intuition about them." Unlike long-distance travelers such as the famous monarch butterfly, Edith's checkerspots are distinct homebodies, occupying habitat patches no larger than several football fields combined. She reasoned that because these butterflies are so geographically constrained, local populations will simply go extinct if their habitat is degraded by significant habitat change, as might occur via climate change.

Four and a half years spent vetting this notion, much of it while living out of her car in a tent all the way from Baja to Banff, showed that 80 percent of checkerspot populations at the southern end of their range in Mexico and southern California had already perished. Parmesan's landmark 1996 paper in the British science journal *Nature* was among the first definitive looks at the effects of climate change on a living creature. The study clearly documented that at least one species was already being affected by global warming; it spurred a wealth of research to assess the effects of climate change on Earth's other plants and animals.

Later, Parmesan and colleagues analyzed many of the new biological studies she had helped inspire. Combing the literature and subjecting data on nearly 1,700 wild species of all kinds to stringent criteria, Parmesan presented strong evidence, again in *Nature,* that some 52 percent of them display signs of having been affected by climate change. Their most rigorous analysis of ninety-nine species in North America and Europe showed that the ranges of a variety of birds, butterflies, and alpine herbs had shifted northward an average of 3.8 miles, or to higher altitudes at an average of about twenty feet, per decade. Scrutiny of data for a truly robust assemblage of 677 plant and animal species in the literature over spans of 16 to 132 years showed worrisome effects: 27 percent showed no change in phenologies (i.e., seasonal timing of annual aspects of life histories such as migration and breeding), 9 percent displayed trends toward delayed spring events, and an overwhelming and highly statistically significant 62 percent demonstrated trends toward spring advancement of breeding or blooming.

This simultaneously brute-force and elegant analysis showed as definitively as possible that climate change is changing the patterns of dispersion of our biota across the face of the planet. Fish, however, were poorly represented in their database, with no diadromous fishes included. Given the weight of the evidence for other animals, we would expect diadromous fishes to have responded similarly, but it's also possible that the flexibility of their life histories somehow absorbed most of the expected effects. Others have since teased out climatic signals on some anadromous species, and it appears that they are no less vulnerable to these forces.

In Maine, researchers quantified the median capture date of anadromous fish—that is, the date when half the individuals in a population that would eventually be seen that year had actually arrived. Their results were remarkable: This date advanced by 1.3 days *per year* for Atlantic salmon in the Penobscot River between 1986 and 2001; and for alewife in the Androscoggin River, by an almost equivalent 1.2 days annually, between 1983 and 2001. Likewise, Atlantic salmon moved up their passage into the

Connecticut River watershed by five days per decade over twenty-three years. Similar findings were made by researchers of alewives in small streams in eastern Connecticut using a different metric. They found the initiation of the run, which they defined as 5 percent of a total run, migrated up their rivers when water temperatures reached 41 degrees, and that 55 degrees best characterized the peak migration. When they looked at how these key variables shifted over some thirty years, the initiation of the runs was now averaging about 13 days earlier, and the peak about 12 days earlier.

Although less studied, the outmigration from rivers to the sea by young-of-the-year anadromous fishes likely also has accelerated. This may be equally as pronounced as in-migration by adults: In the Loire River, France, the first day when 5 percent of the catches of Allis shad were made was seventeen days earlier in 2004 than in 1995.

Sea lamprey on spawning site, Basher Kill, New York
John Waldman

What does it mean to shift phenologies ahead by about two weeks in fewer than three decades? And did temperatures in rivers change much but go unnoticed in the preceding decades to a century or more before these analyses? There is flexibility within normal migration schedules for anadromous species because in typically abundant populations, individuals arrive over a period of weeks to two or more months. But this window has constraints, too. A fish entering a river too early usually encounters colder and faster-flowing waters, but the lower temperatures force lower swimming capacities for these cold-blooded animals, and with more energy needed to be expended fighting the faster currents. On the far side of the season, a fish entering a river should enjoy slower currents, but possibly too little water to reach spawning grounds. And higher temperatures create a tougher physiological slog requiring greater energy expenditures.

With these countervailing forces there has been long-term selection for optimal timing of migration for each species in each watershed. It seems probable that the rapid changes already documented for the few examples examined so far represent only a partial adjustment to the additional temperature changes to come (and to those that already occurred and went undocumented), a rapid and radical temporal restructuring of migrations that may exceed the ability of the species to adapt on the fly (or the swim). But beyond the shifting migration clock is the high likelihood of disengagement from the production cycles that nourish the young fish spawned. Once the newly hatched "fry" use up their yolk sac, they must make a living from the prey available in their river or estuary. The migrations, spawning times, and development times of the eggs and fry of anadromous fishes are fine-tuned to the ecologies of their watersheds, and not all components of their food webs will necessarily move their seasonal clocks up in lockstep.

This is especially true for phytoplankton and plants, which respond to changes in the daily availability of sunlight (which doesn't change) instead of to temperature (which does). Thus, warming may induce mismatches of fish reproduction to the quickly evolving ecological framework that supports it. And smaller cohorts produced in a given year yield smaller numbers of adults returning and, consequently, a shorter spawning run, meaning that at least some component of the young produced are less likely to "match" the spring peak of prey production. It's easy to see how these circumstances could result in a spiral downwards for a population.

Regardless of the reality of postindustrial climate change, the Earth's atmospheric temperatures have varied naturally on many other scales, from tens to millions of years. A controversy was spawned with the conjecture by Catherine Carlson, then a doctoral student at the University of Massachusetts, that Atlantic salmon were not the abundant long-term inhabitants of New England rivers that was commonly assumed. She described this widely held belief as folklore, citing examples such as the oft-quoted story that early colonists could walk across the fishes' backs in rivers. Scholars had also cited their apparent abundance, with one stating that "the Atlantic salmon rivaled the cod as an important and reliable source of protein to the early New England colonists." And another claimed that there is theoretical reason for thinking that Atlantic salmon, per unit area, was at least as plentiful as Pacific salmon.

But Carlson argued otherwise. She noted that the remarkably complex and dependent societies of the Pacific Northwest based around salmon fishing were impacted so much later than any that might have existed in the New England region that they were well described by ethnographers as late as the turn of the twentieth century. But on the Atlantic Seaboard, the introduction of European diseases was so quickly devastating to the aboriginal peoples that little evidence of the importance of these fishes and to their abundance remains except through archaeology. And to her mind, the archaeology simply was not supportive of great salmon abundances.

In 1980 Carlson began a study of the prehistoric fisheries of Maine's Boothbay region through archaeological analysis of fish bones excavated from ancient middens at twenty-one sites. Surprisingly, fish bones had never been analyzed for New England (such talents being an uncommon specialty in zooarchaeology). Her initial examination of some 30,000 bones revealed not a single salmon element, even though one of the sites was along the Sheepscot, a small river that historically supported salmon.

This unexpected finding sparked Carlson to expand her reach, soon showing that there were no skeletal remains of salmon in site after site in New England, although bones of many other fish species were recovered. Eventually, her review of bone remains from more than seventy-five New England locations revealed only four possible of salmon—two vertebrae each from two locations—and all instead may be from the morphologically similar brook trout. Carlson wondered how this could be. Did aboriginal peoples not catch them because they lacked suitable gear? Or because they did not like to eat them? Did the bones not survive in the soil conditions of New England? Were the historical accounts grossly embellished fish tales? Did the salmon runs not exist in New England, or were they so minimal as to be undetected archaeologically?

Clearly, the archaeological and historical records clashed. The possibility that Native Americans could not catch salmon in rivers or that they found them disagreeable could not be supported. That the bones do not survive New England's soils was discredited by the equally fragile bones of many other fish species having been preserved; moreover, salmon remains at archaeological sites in the Pacific Northwest have persisted in great quantities.

Carlson then turned her attention to critically reviewing the primary historical records about New England fishes. She found that sources claiming great quantities of salmon were nineteenth- and twentieth-century syntheses and compilations written long after the fishing occurred, and often were based on hearsay. Two examples were particularly egregious. One was the unreferenced statement in a book on salmon that in colonial New England, salmon "were sometimes so thick in the rivers that they overturned small boats." Likewise, another for Lake Ontario, that salmon were once so numerous that women "seined them with flannel petticoats." This debunking was taken further by another investigator who reviewed the old story concerning the Connecticut River, of the apprentice agreements which were supposed to have protected the poor with a clause stipulating that they not be required to eat salmon more than twice a week. He wrote: "As a matter of fact, it is an English or Scottish tradition which is not true, even in the land of its origin. As long ago as 1867 the London Field offered a reward of five pounds to anyone who could produce one of these agreements. The reward was withdrawn a year later, unclaimed."

To gain a clearer read on the issue of salmon abundances, Carlson surveyed the primary historical documents of the seventeenth and eighteenth centuries, still within the time prior to the extensive dam construction widely considered responsible for the salmon's decline. Because these accounts did not provide actual figures on salmon numbers, the statements concerning salmon were compared with those for other fish species in order to gain a general impression of *relative* abundances. These fragments of evidence by explorers, merchants, and travelers together indicated that some salmon were present historically; they do not support the widespread notion that they were wildly abundant. She found that when species of fish are listed or described, salmon, if mentioned at all, tend to fall toward the middle or end of a species list, suggesting their lesser significance. Some accounts go into considerable detail in describing particular fish species, but are much more cursory in their references to salmon. And a number of sources did not even mention salmon. It was not even necessarily well respected where it was available; a 1749 account reported that the salmon of the Merrimack and Connecticut Rivers "is not of a good quantity and is not so good quality and is not so good for a market as the salmon of Great Britain and Ireland."

Even if the conclusion of Carlson's review of the historical literature on salmon is accepted—that salmon were only a minor resource—it doesn't explain the virtual absence of the fish in the archaeological record. One unanticipated but apparent resolution is that the presence and absence and numbers of salmon have actually varied significantly over recent centuries. Carlson hypothesized that salmon did not colonize New England streams until the historic period. Atlantic salmon is a coldwater species, and New England is at the southern extreme of its range. Perhaps a favorable period of climatic cooling known as the Little Ice Age, lasting from about AD 1550 to 1800, created sustained conditions cool enough to allow a shift in range to lower latitudes. Salmon might have suddenly established at least marginal populations in a number of rivers from Maine to as far south as the Housatonic River in western Connecticut. Under this climatic-based scenario, warming at the end of this period would have created less-favorable conditions, causing the salmon's range to retract naturally. Accept this theory, and you explain the lack of salmon in prehistoric sites, the apparent limited abundances of salmon historically, and the extirpations from many New England rivers, especially more southerly ones, at the end of the eighteenth century.

Carlson, of course, needed to account for the source of these new salmon colonists at the beginning of the Little Ice Age. The story gets a bit sketchy here, though she does assemble a plausible scenario in which salmon migrated from Europe after the end of the Pleistocene, perhaps with a warm period between AD 900 and 1300, known as the Little Climatic Optimum, melting sea ice and allowing them to reach the Davis Straits between Labrador and Greenland, and then moving southward from Labrador with the Little Ice Age. Perhaps . . . but maybe not.

To her credit, Carlson, with her provocative doctoral thesis, initiated a cottage industry of researchers who have tested her theses. Some have been found wanting, while others have held up. An eclectic group of researchers at the University of Maine led by Brian Robinson reviewed Carlson's work and other evidence and concluded that her case was not as strong as presented. Only twenty of the seventy-five sites she included in her study were at the inland locations where one would expect to see salmon remains. But reconsideration of known materials and new sites now show salmon material from seven locations in Maine, and they are not all recent; two date back approximately 6,000 years.

If salmon were unambiguously present and harvested, why then are their remains so rare across thousands of years of aboriginal presence? Robinson and his colleagues point out that salmon were harvested only at selected, strategic sites, and that some of these may have been inundated with rising post-glacial sea levels; that fish remains need to be exposed to fire in order to enhance their preservation in the acidic soils of New England; and that the fragile bones of fish seem to have survived only from places where rapid burial occurred. Shallow sites often have abundant mammal but little or no fish remains.

Carlson's conclusion that Atlantic salmon are only recent arrivals to North American rivers is contradicted by other kinds of evidence. More likely, in my view, is that salmon were present during the Pleistocene in then-chilled rivers just south of the ice sheets, on the simultaneously exposed Georges Bank offshore of New England, or both, as genetic results suggests occurred for some other anadromous fishes that used unglaciated rivers as refugia during the ice ages. Physical evidence for this for salmon was lacking, though, until climate investigators plumbing the sediments of a small pond in northwest New Jersey examined the cores they'd drilled and discovered seventeen fish scales, among them what appears equivocally to be a single salmon scale that may confirm their presence in the Delaware River drainage under colder conditions, some twelve centuries ago. Also, European colonists found established populations of "landlocked" salmon in four large inland lakes in Maine. It is doubtful that salmon could have reached these disparate lakes and formed ongoing populations in a brief period just before the Europeans landed. Finally, colleagues and I recently sequenced salmon DNA from both continents and found that when a "molecular clock" or standard evolutionary rate was applied to the degree of divergence between salmon from both sides of the Atlantic, it indicated a time of separation of between 1 and 2 million years, meaning that Atlantic salmon have swum in North American rivers for quite some time.

The Atlantic salmon controversy, regardless of the exact details, does suggest a high sensitivity and responsiveness by anadromous fishes to changing climatic conditions. Today we know that Atlantic salmon are feeling the pull on their phenologies for spawning adults to migrate up rivers earlier in the season. But how salmon reacted to temperature in the Little Ice Age would have constituted the usual fine-tuning of abundance and distribution to normal climatic forcing. What's occurring today climatologically appears swift, unidirectional, and, unfortunately, unstoppable in comparison.

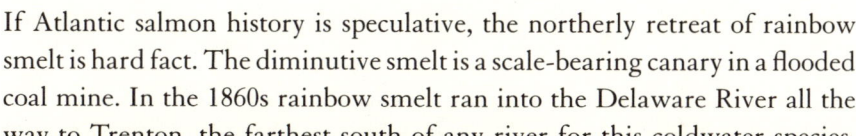

If Atlantic salmon history is speculative, the northerly retreat of rainbow smelt is hard fact. The diminutive smelt is a scale-bearing canary in a flooded coal mine. In the 1860s rainbow smelt ran into the Delaware River all the way to Trenton, the farthest south of any river for this coldwater species.

But their demise there was little noticed, and a latitudinal draw-back began. Commercial catches were reported from the southern tip of New Jersey at Cape May County until 1904, and the last commercial catch in Middlesex County in central New Jersey in 1921. Farther north, in Newark Bay, a now royally polluted arm of the Hudson Estuary, they were caught by wagon-loads in the 1880s. In the Hudson itself, smelt ran in catchable numbers into tributary streams in February and March when the river reached a chilly 40 degrees—but not after 1979. Scientific monitoring showed that for a time they did still spawn in the main-stem river, but the last few specimens were seen in 1998.

The Hudson Estuary was only a way station on the smelt's line of retreat. Just east of it many Connecticut streams and brooks enter Long Island Sound, with most enjoying smelt runs through recorded history. But they also dwindled and disappeared late in the twentieth century, again with little notice. In 2003 and 2004 researchers at the University of Connecticut intensively surveyed known smelt-spawning and smelt-fishing rivers along the entire breadth of the state, using many kinds of gear meant to catch every life stage from eggs to adults, and just nine adults were caught, from a seine net in the Mystic River in the far eastern part of Connecticut.

Where do warming waters leave the smelt? In 2004 this little fish that once ran up rivers in untold numbers in New York and southern New England was named as a Federal Species of Concern—a foothold, per-haps, as an eventual Endangered Species listing? There is little doubt that climate change is the primary cause. The normal fluctuations in climate on the order of decades show particularly high sensitivity of a species to temperature *at the edge of its range*. Between 1920 and 1930 in Connecticut, temperatures declined and smelt catches rose; a warming episode occurred between 1930 and 1938, and catches fell. During a winter cooling period in the 1960s, commercial activity for smelt in Long Island Sound recovered briefly. Indeed, I well remember rushing from junior high school in late November in the 1960s to squeeze in an hour of smelt angling before dark off a dock on the Bronx shore of the Sound. A warming trend in the 1970s brought the Sound's commercial smelt fishery to a halt; not knowing that, I returned there in the late 1970s, hoping to relive that memory, but the smelt were gone—likely forever.

Is every change in a fish's distribution attributable to warming? It was August 1997 as I drove my family up a ramp onto the ferry that takes vacationers and residents from Black Harbor, New Brunswick, to Grand Manan, a twenty-one-mile-long island near the western side of the mouth of the Bay of Fundy. The hazy sunlight of the mainland shore gave way to a chilly fog as we made the crossing for the first time. We had no idea that striped bass, a species that lords over the coastal Atlantic, had also just made that same crossing for its first time, at least in recorded history.

As an ardent angler, I asked around about fishing opportunities and was told that if I was to go to the docks at the marina, catch a small pollock, and put it on a larger hook and let it swim around, I would quickly catch this exotic fish called a striper. Sage advice; my son Steve and I quickly landed a few. But more interesting to me was the novelty of their appearance to the islanders. Fishermen told me that these fish had shown up the year before, and many of the locals had had no idea what they were, having never seen them in their waters.

A few days later as we were walking along a pier to board a whale-watching vessel at a different marina, I looked in the clear shallows along a beach and was startled to see a school of about seventy-five striped bass moving lazily, much like a weather system that circles at the same time as the entire mass drifts in one direction. Clearly, many stripers had reached the island, but why only now had they swum the nine-mile-wide channel through 400-foot depths to get there? Striped bass from American rivers were known to reach the Bay of Fundy to the north, but they had always avoided Grand Manan Island.

In 2007 I attended a conference in Halifax, Nova Scotia, and then took a couple of days to explore the province's spectacular north end, Cape Breton, in particular, a unique body of water I was intrigued with, the Bras d'Or. The Bras d'Or is a sprawling 425-square-mile inland sea of moderate salinity, but only minuscule tides because of its having only two narrow connections with the ocean. Its depths had long contained true but largely landlocked marine species such as cod, pollock, and flounder, but never striped bass. Decades earlier I had read a now long-lost magazine article about catching small cod on light tackle from shore, and I set out to do the same. But sea fishing in the Canadian Maritimes is still mostly a commercial endeavor; sportfishing equals trout and salmon, with rod and reel along the coast still in its embryonic stage, hardly a sea angler to be seen, and information nearly nonexistent. Somehow, though, inquiries led me to a causeway on the east side of the Bras d'Or, where two hours of casting produced zero cod. Then a local stopped his car and said cod could be caught there, but only at sunset. Because dusk was still nine hours away, I went exploring and never made it back to this spot.

In 2009, back in my office in New York City, I received an e-mail from an angler who had just landed a giant striped bass from the very same causeway, one that qualified for the Nova Scotia province record. Moreover, it wasn't the only one. That year catchable numbers of mostly very large striped bass had invaded the Bras d'Or, the first time anyone knew of the species there.

Were these new occurrences of striped bass around Grand Manan Island and in the Bras d'Or the result of climatic warming, or something else? After all, the close evolutionary cousin of striped bass, the European sea bass, had greatly expanded its range beyond England in the United Kingdom as its water temperatures rose in recent decades, suddenly becoming common in Scotland, and even making a northward jump to Norway. All the unseen influences and shifting factors in the marine world that affect creatures as complicated as fish make it impossible to unwind confounding factors with surety, but my hunch is that the sudden appearance of the stripers at Grand Manan Island had nothing to do with warming, while their appearance in the Bras d'Or had everything to do with warming.

Populations of fish exhibit something that scientists call "density-dependent distributions"—that is, the idea that the abundance of a population affects its coastal range because, simply put, for individuals in the population to minimize competition with each other when a population becomes large, they must spread out farther. The shrinking and expansion of a population's range can reveal its most preferred habitats. When a population is small, it can be choosy and mainly occupy its most suitable territories. When it expands, it begins to inhabit areas of lesser desirability so as to reduce the within-species competition. This may be shown both by occurring in less-suitable habitats within its normal range, but also through expansion of its coastal range; in the case of striped bass, this was seasonally northward, because summer temperatures are favorable.

The novel appearance of striped bass at Grand Manan Island took place at the very peak of the striped bass recovery, a time when northern locations like Maine enjoyed unprecedented striped bass angling. Simple population pressure probably drove some northward-migrating schools to adopt more-exploratory behavior.

The striped bass that surprised anglers in the Bras d'Or were not numerous and were drawn from stock sizes that had dropped considerably since the mid-1990s; indeed, many American anglers were pleading for new restrictions on harvests because they saw history repeating itself, with another striper crash on its way. More likely, these fish are now routinely making more northerly summer migrations as waters warm, much like the European sea bass did, and, in the course of following the coastline, were drawn to the warmish and fertile waters of the giant inland lake. Like the times, the seas are a-changin'.

Recently, a couple of sharp-eyed data crunchers found that one sea (actually, a very large estuary) has been changing according to a regular cycle, one so pronounced that at least for the Chesapeake, we should be able to predict the fates of some anadromous species over a decade or more. They proudly named it CBASS, standing for Chesapeake Bay Anadromous and Shelf-Spawning Species. Robert Wood, then a doctoral student at the Virginia Institute of Marine Sciences working under Herbert Austin, amassed long-term data on fifteen fishes from four independent surveys that spanned from 1968 to 2004. The fishes included both anadromous species and those that spawn on the coastal continental shelf, like menhaden, spot, bay anchovy, and summer flounder. Wood and Austin's insight was that although the anadromous fish spawn in rivers and the marine species spawn on the continental shelf, in both cases their young spend much of their early existence in the same estuaries.

Some fancy statistics showed a strong pattern: When, as a group, anadromous species showed high abundances of young, the marine species did not—and vice versa. In fact, production of young by fish is notoriously variable on a year-by-year basis. Yet the investigators found that whereas 38 percent of the total variation occurred annually, an amazing 62 percent varied on a decadal scale. Clearly Wood and Austin were on to something. But what controlled this newly revealed flip between these two states? Water, the master variable, of course. When freshwater runoff is high for extended periods, estuaries tend to have lower salinities that favor the freshwater-spawned anadromous fishes. And when runoff is lower, the marine species flourish. But this begs another key question: Why would freshwater runoff vary on a scale of several decades or so? Wood came to believe that this is a continental effect of the Atlantic Multidecadal Oscillation (think a small El Niño, as in the Pacific). The warmer of its two states features a high-pressure system over the Ohio Valley that blocks storms from moving quickly up the coast, which dumps more rain and snow on the Mid-Atlantic.

Wood says we recently flipped into the drier state, and that anadromous fish will be spawned in lower numbers until it reverses a decade or so from now. Also, the indexes of striped bass production in Chesapeake Bay were low between 2008 and 2010. But remember that 38 percent of the variation occurred between years; 2011 experienced abundant rain and especially heavy snowmelt following record-breaking winter storms. There was lots of freshwater in the estuary, and by the end of summer, biologists knew there were lots of little stripers, too, the fourth-highest index over the survey's fifty-eight years. Then came the spring of 2012, with drought and high temperatures following a warm, dry winter. CBASS would predict low striped bass production. And low it was—a record across the almost six decades.

Even as these two categories of fishes compete to dwell in estuaries, the total expanse of these critical habitats will shrink. Prehistorically, estuaries and wetlands simply tracked sea-level changes, on the scale of tens or hundreds of feet as liquid water became more or less available with the growth and shrinkage of the great ice sheets. But today, as we've built homes and other structures up to the edge of estuaries, there is nowhere for them to track inland as sea levels rise. Sea levels along the Eastern Seaboard between Boston and Cape Hatteras have been rising at 0.8 to 1.5 inches per decade since 1950, three to four times the global average. Furthermore, they are projected to rise as much as nearly two feet or more by 2100, though there are credible doomsday scenarios that could dwarf that.

It is said that "Climate is what you expect; weather is what you get." But climate varies naturally on many scales, from across hundreds of thousands of years to the centennial, or even less. The standard averaging period for climate is thirty years of "weather." Twelve of the hottest years on record, worldwide, occurred in the last fifteen years. There is no credible doubt now that the world is steadily heating up due to humankind altering the Earth's atmosphere. The most recent report of the relentlessly sober Intergovernmental Panel on Climate Change gives predicted worldwide air temperature increases by the end of the twenty-first century that range from 2 to 11 degrees Fahrenheit, depending on how well humankind does in reining in atmospheric emissions. Frightening changes. Time may soon tell how many of our aquatic creatures truly are in hot water.

Chapter 13

Migration and the Exotic Species Gauntlet

> You would think that time would be more compatible
> with the tide—time and tide, this daily up and down—
> but somehow, I think there is a lot to be learned about
> time by the river.
>
> —*Andy Goldsworthy,* Rivers and Tides

David Strayer saw a river changing before his eyes. A remarkably productive researcher at a rarefied establishment, the Cary Institute, Strayer was one of a cadre of young Institute scientists who'd made a long-term commitment to understanding the entire Hudson River food web—a look into the machinery of how the river "works" at its most fundamental levels, from nutrients through fish life. What he saw worried him; the exotic zebra mussel was hijacking the river in ways that would ripple through its ecology, including its diadromous fishes.

Invasive species, *exotic* species, *alien* species, *nonnatives:* These are somewhat interchangeable terms that differ in their shadings. At the extremes, invasive species usually means those that are problematic; nonnatives may (or may not) be perceived to be benign. Regardless of the nomenclature, the movements and colonizations by exotic species around the globe are a major conservation problem, an environmental version of the rampant globalization occurring in societal form. Exotic species can disrupt existing food chains by preying upon or outcompeting native species. And, unless they are recognized almost immediately when the first pioneers appear in their new habitat, exotic species can rarely be eradicated.

Exotics arrive in new places by many avenues, called *vectors* in conservation biology. Thankfully, a once main and highly direct vector is declining in importance—the purposeful introduction of new species. More than a century ago, "playing God" in this way was viewed as a kind of hobby for Victorian dilettantes. Some belonged to "naturalization societies" that sought to increase the breadth of species in their bailiwicks by naturalizing them to their new conditions, with almost no concern for how the new animals and plants would compete with or harm the existing biota. In the United States the federal government and state conservation departments took this further, "trying" new desirable species in their territories, as if they could just be removed if they didn't fit in, in the ways imagined.

It's true that some introductions do play useful—although usually somewhat controversial—roles. Before Appalachian and New England streams were dammed and their banks deforested, they ran cool and were home to great abundances of native brook trout, a temperate member of the mostly boreal chars. But such changes rendered these watersheds less suitable for brook trout, and they retreated to the colder headwaters of some systems and disappeared altogether from others. Stocked Eurasian brown trout—first introduced in 1883 in Michigan—filled the void. Yes, brown trout are not resplendent like the colorful brookies, and yes, they are harder to catch, and yes, they may outcompete and even eat brook trout where they overlap, but many thermally altered American rivers would not offer the pleasures of trout fishing without them. There are anglers who wish the browns hadn't crossed the Atlantic, but they are vastly outnumbered by those who give thanks.

Then there are introductions like the common carp, a fish whose aficionados are swamped by its detractors. There is no doubt this European and Asian species now flourishes in every US state except Alaska; how they reached America is still debated. The naturalist James Ellsworth DeKay received a letter from a Henry Robinson who said he brought some then-esteemed carp to his ponds from France in 1831 and 1832, and that he released some to the Hudson River. But Spencer Baird of the US Fish Commission examined Hudson River specimens and declared they were goldfish or goldfish-carp hybrids. Others claimed that carp entered the continent through Connecticut and California. Regardless, these giant "minnows" that have exceeded 100 pounds soon came to dominate many water bodies where they grub through the bottom, eating the eggs of and outcompeting more desirable native fishes and mucking up the water column, allowing less light for photosynthesis of ecologically productive weed beds. Purposeful introductions, misguided or not, are only one of many vectors for nonnative species.

"Bait bucket" introductions are also common. Wild minnows may be caught somewhere and distributed for sale at bait shops. Mixed among the hundreds are a few juveniles of other species of the same sizes. Or maybe the wild or aquacultured bait minnows themselves aren't native to where they are being fished. An angler buys bait, doesn't use it all, and then doesn't have the heart to kill what's left, and so dumps them where he or she is fishing. Suddenly, an alien appears in the mix, no doubt influencing the ecology of that water body in some fashion—a predator, competitor, or both.

Fish can also simply swim to where they don't belong. Before there were highways, goods went back and forth between the interior and the coast of many middle-Atlantic and northern states via man-made canals that linked formerly isolated watersheds. Many of these canals are defunct and some have been filled in, but they were far more numerous than one might suppose based on the few now mostly recreational canals that have survived. Not every fish species would utilize the mostly still habitats of these waterways, but it appears many did. The Hudson River, for example, is still coupled with two major thoroughfares: the Champlain Canal, connecting it to that almost Great Lake, and the Erie-Mohawk system that linked it to Lakes Ontario and Erie. At least fourteen fish species made it to the Hudson from the Midwest through canals. Now, in 2013, a large European carp-like fish, the tench, is poised to enter the Hudson through the Champlain Canal.

The list of vectors goes on and on. Aquaculture operations can "lose" fish in storms that overwhelm facilities or through carelessness when escapes occur. A truly awful example of this occurred across the Atlantic in France, where the last European sea sturgeons survive as a relict population spawning in the Gironde River. French biologists were nervously trying to maintain this stock through captive breeding and releases of cultured young, but they were having poor success producing juveniles, and some of the scarce breeders were dying in the process. To avoid losing more sea sturgeon, they brought a nonnative surrogate, Siberian sturgeon, to their culture facilities near Bordeaux to experiment with, toward improving their sturgeon husbandry.

Then in 1999 came the "Storm of the Century," with 125-mile-per-hour winds, the breaching of dikes, and the flooding of twenty square miles of lowlands, including a fish farm harboring 67 tons of Siberian sturgeon. When the waters receded, only 27 tons were left, which translated to about 9,000 missing fish. During the next few weeks, 1,045 Siberian sturgeon were gathered up by farmers and fisheries workers in marshes and ditches around the fish farm. This meant that about 7,900 sturgeon washed into the river, including 2,400 ready-to-spawn adults. The best outcome would be that the Siberian sturgeon simply coexist with the European sea sturgeon for a while and don't reproduce, but it's likely they will compete directly with the sea sturgeon and could even hybridize with them, which would essentially breed them out of existence in the wild. Likewise, the Pacific pink salmon being aquacultured in Russia escaped, invaded several rivers in Norway, and is now threatening to outcompete the far more valuable native Atlantic salmon.

Invertebrates and plants also travel to new watersheds inside and outside of boats. Some may have life stages that cling surreptitiously to boat hulls. Others take a ride in the ballast waters of large commercial vessels. A ship leaving a fresh- or brackish-water port in Europe may take up vast volumes of water there to stabilize an Atlantic crossing, only to discharge part or all of that water when it picks up additional cargo on the other side. Ballast tanks are like large, dark lakes that allow many mostly young life stages of plants and aquatic insects and other invertebrates to arrive at far distant destinations. This is how the zebra mussel invaded the United States.

Somewhere around 1985 zebra mussels hitched a ride in a ship from a European port and up the St. Lawrence Seaway to Lake St. Clair, near Detroit. From there they quickly spread through the Great Lakes and beyond. Zebra mussels were first seen in the Hudson near Catskill, New York, in May 1991. Catskill is an area that hosts numerous black bass fishing tournaments, and it's possible that the mussels were introduced by a fisherman who trailered his boat from competition to competition. To say the mussels thrived in the soupy Hudson is an understatement; by the end of 1992 David Strayer estimated the river contained 550 billion of them!

These bivalves, which coated hard submerged surfaces such as rocks and sunken logs, filtered roughly the entire volume of the river each day. Such a scrubbing had profound effects on the flow of organic matter and dissolved nutrients in the river, which rippled up the food chain. Most easily observed was that the Hudson's water became somewhat clearer—not dramatically so, as in eastern Lake Ontario and the Thousand Islands region, where the waters became downright pellucid, but enough so that sunlight penetrated farther, allowing weed beds to grow deeper and extend from shore.

Strayer wanted to quantify the changes he believed he was witnessing to the Hudson's fish life, and so he and colleagues capitalized on the superb long-term database on Hudson River fish populations supported by the power utility companies on the river. Long-term monitoring of rivers is expensive and scientifically boring (at first), but it *always* is just a matter of time till it proves worthwhile, if not even essential. Comparing data from before and after the zebra mussel invasion, they found that vegetation-loving fish such as black bass and sunfish became almost doubly numerous, grew faster, and had shifted the center of their distribution upriver toward purely freshwater reaches where the weed beds grew most luxuriously.

But many of the anadromous fishes had the opposite response. The mussels reduced phytoplankton abundance in some river reaches by as much an incredible 80 percent, and zooplankton biomass by 50 percent. This might appear terminal to any creatures that depended on them, but the Hudson has a parallel food chain based on the bacteria that consume the detritus that washes off the watershed, and this helped to buffer the phytoplankton losses. Nonetheless, when Strayer and his colleagues examined data for striped bass, shad, and river herring, they found that they had become statistically significantly less abundant, grew slower, and had shifted their center of distribution downriver toward saltier regions, where zebra mussels have less effect and the plankton they feed on were denser. One of the great mother rivers of anadromous fishes to the Atlantic Ocean was being compromised by the lack of preventive attention given to tiny bivalve larvae in the hold of at least one ship a decade earlier.

Not all nonnative species listed in scientific papers and reports for a given watershed have meaningful ecological effects. Ichthyologists are an obsessive lot, prone to keep track with checklists that even include single catches of species unlikely to occur again, such as releases of tropical-fish pets that are doomed to perish in chilly temperate winters. Nonetheless, perhaps no Atlantic coast river is now free from exotic fish with the potential to prey on native diadromous fish. And in some rivers these exotics become a formidable gauntlet for young-of-the-year outmigrating anadromous fish to have to swim through. Other exotic fish don't prey directly on diadromous species but do alter the ecology of rivers in substantial ways.

In northern Atlantic coast rivers that had been glaciated during the Pleistocene, fish diversity was naturally low given the challenges in recolonizing these newly exposed waters from refugia in less than 15,000 years. In many of these rivers the predatory suite of fishes that would take young anadromous fish moving down to the sea included only chain pickerel, the relatively diminutive yellow perch and white perch, and any striped bass that had moved into freshwater. The anadromous species had evolved over time to cope with this low-caliber assemblage, but then humans began to add larger and more-aggressive species to the mix.

Maine is beyond the natural range of both important East Coast black basses, largemouth bass and smallmouth bass, but they were introduced early—1868 for the smallmouth, and likely in the late 1800s for the largemouths. Yet Maine is highly regarded for its black bass fishing; Spednic Lake, in the St. Croix River system, is thought to offer the finest smallmouth fishing anywhere in North America. A black bass is a vacuum cleaner with fins; they can expand their copious jaws and inhale (using suction) large prey such as frogs, mice, and crayfish, not to mention plenty of outmigrating shads and river herring. This voracity plus their large size and fighting abilities make them revered by anglers. Yet these bass exist in a kind of reversed shifted baseline. Because they were caught by your father and your grandfather, and probably a generation or two before that, they often are assumed to be native in the Northeast, i.e., our grandfathers have grandfathered them in. No one is advocating black bass eradication.

In the Hudson River there has been a 130 percent increase in the number of exotic fish species reported since the 1930s, and some were there well before that. Carp and rock bass were noted in 1842. Those that followed include predatory types such as northern pike, hybrids of northern pike and muskellunge, largemouth bass, smallmouth bass, channel catfish, freshwater drum, black crappie, and walleye—a truly formidable array of jaws to have to swim past in what had been the comparatively safe nursery haven of fresh waters. Introduced goldfish that reach a couple of pounds and the recently naturally invading gizzard shad, together with the common carp, all flourish in the Hudson, and now form a substantial portion of the total fish biomass of the river, changing its food chain in unknown ways, but almost certainly not in directions favorable to diadromous fishes.

Yet one more species was a close call for the Hudson, having been found in a tributary to the river in 2008 and quickly eradicated. The northern snakehead is a voracious Asian fish that was imported to the United States for consumption by Asian communities here. Tougher than nails, not to mention bolts and screws, it thrives in swampy waters and periodically takes air from the surface. In 2002 snakeheads were found in three ponds in Crofton, Maryland. That other snakehead species are known to wriggle across land, that they are primitive-looking, and that they are toothy and voracious somehow caught the media's and the public's attention. They became "Frankenfish," and were even joked about on the *Late Show with David Letterman*. The response to the poisoning of the lakes with rotenone, a fish-killing chemical, became such a circus that opportunists sold commemorative T-shirts. It was surmised that the snakeheads had been in the lakes for only two years, but their populations had exploded—more than a thousand young and six adults were recovered.

One of the Crofton ponds was only seventy-five feet from the Little Patuxent River, which would have given them access to all of Chesapeake

Bay. Though the eradication was considered a success, they did show up in a major tributary to Chesapeake Bay, the Potomac River, in 2004. Today an angler can target snakeheads in the Potomac and fully expect to catch some. Your great-great-great-grandchildren can expect the same someday.

It is safe to assume that every East Coast river has a few to a dozen or more exotic fishes that directly or indirectly influence the native diadromous fish community. The Susquehanna has sixty native fishes and more than half as many more nonnatives, the aliens including blue tilapia that escaped from an aquaculture facility and the highly carnivorous flathead catfish, a wide-mouthed monster that elsewhere has reached about 125 pounds. They are joined in the Chesapeake system by a highly predatory pair, channel catfish and another midwestern behemoth, blue catfish. Flatheads also recently showed up in the Delaware, something that will be taken as bad news by its many anglers dedicated to its (nonnative) trout.

The trend to a high number of exotic fishes in rivers continues latitudinally all the way to Florida, only their identities slant toward the tropical. The smallish Kissimmee River near Orlando, Florida, has only eels among diadromous species, but it has a suite of other fish that don't belong there that includes the same common carp found up north, but also the Asian grass carp, a weed eater that reaches forty pounds, the infamous walking catfish that can waddle across wet grass, a South American armored catfish, blue tilapia, and a tropical aquarium species, the oscar.

On the horizon is a threat at least as worrisome as snakeheads: Asian carps are knocking on the door of the Great Lakes, if they are not in there already. Bighead carp, which can reach 100 pounds, and silver carp, which top 60, eat low in the food chain, which made them attractive as "work" fishes to keep algae at bay in aquaculture and waste-treatment facilities. But they escaped to the Mississippi in the 1970s, gaining access, in theory, to most of Middle America. Recently they reached the Illinois River and are being stopped from passing into Lake Michigan only by an electric barrier in an important shipping canal. A contentious state-of-the-art DNA-based approach says they're already in the Great Lake. Why is this troubling? One reason is that the silver carp have a wacky instinctual urge to leap as high as ten feet out of the water when they sense a motorboat engine. This has resulted in numerous and sometimes serious injuries of boaters and jet ski-ers who collide with them while speeding along; in 2003 a woman jet skier broke her nose and vertebra and almost drowned after being struck by one. And forget waterskiing. The other reason is that they thrive in nutrient-rich waters, as in parts of the Great Lakes and many East Coast rivers. If they colonize Lake Michigan they will eventually reach the St. Lawrence and the Hudson, where they could constitute much of the piscine biomass of these rivers, leading to ecological havoc.

The Hudson and some other East Coast rivers host another noxious invader, the water chestnut plant *Trapa natans*. Though Asian, this is not the water chestnut one sees in Chinese cuisine; instead, it is a plant that has broad floating leaves that cover tens to hundreds of acres of mucky-bottomed bays in the river. The species was introduced into North America by well-meaning but naive botanists, one of whom wrote "but that so fine a plant as this, with its handsome leafy rosettes, and edible nuts, which would, if common, be as attractive to boys as hickory nuts now are, can ever become a nuisance." But it became a damned nuisance. Deliberately introduced to a lake in upstate New York in 1884, it escaped into the Mohawk River by the 1920s, and then into the Hudson in the 1930s. Swimmers found that the sea urchin–like spiny seed cases, known locally as "devil horns," are painful to step on. To see water chestnuts in the Hudson is to view a sea of green—flat leaves so tightly interwoven that little sunlight penetrates below. Little light equals little photosynthesis in the water column under the leaves. Dissolved oxygen levels can fall below levels needed for fish, as low as zero in large dense beds. Thus the shallow bays so critical as nursery areas for many fish species may become distinctly unproductive for fish, harming the overall ecological functioning of the river.

Not all exotic species that affect diadromous fishes are predators or competitors; new parasites can suddenly materialize, too. There is little rou-tine screening for fish parasites, but some are just too obvious to miss. A thin red nematode worm that infects the swim bladders of Japanese eels showed up in eels in Germany in 1982, a foothold in Europe that later ballooned to Sweden, England, Portugal, Egypt, and Russia, and in between—pretty much the entire range of the European eel.

Anguillicola crassus, meaning something like "dense" or "gross" of the eel, was first seen in the United States in 1995 at a Texas aquaculture facility, in eels obtained from South Carolina. Examination of thirty wild specimens in South Carolina revealed one with the parasite, showing that the nematodes were established in the wild in the United States. In spring 1997 watermen on the Patuxent River told Chesapeake Bay researchers that the eels they were catching contained "worms." A large survey of the Bay showed prevalences of 10 to almost 30 percent by location, and intensities of as much as two dozen worms in a single eel.

The parasite co-evolved with the Japanese eel, which has some resistance to it. Like many parasites, *Anguillicola crassus* has a complicated life cycle involving intermediate host organisms, but our native zooplankton seem to play that role well, and so the worm is flourishing in the less-resistant American eel, with infections seen from Cape Breton, Nova Scotia, to Florida. And the worm has nasty consequences; intense infections can cause hemorrhagic lesions of the swim bladder, swim-bladder fibrosis, skin ulcers, and swollen anal glands. There also are concerns as to whether eels with a heavy fibrosis of the swim bladder can successfully navigate all the way to their spawning grounds in the Sargasso Sea. The full effects of this parasite colonization are not yet known, but it's quite safe to assume they are not beneficial to a species already in steep decline.

Yes, some nonnative introductions can be viewed as being benign or positive in their effects. Yes, there is ever-growing awareness of the exotic species threat and stronger efforts to reduce it. And even the odds appear to be in favor of an introduction not causing serious problems. Early research has shown that although our defenses against invasives are remarkably leaky, actual successful colonization conforms to the highly approximate "10 percent rule." That is, only about 10 percent of the newcomers actually settle and live in the new waters and, of those, only about 10 percent establish themselves. Onerously, new work by Strayer and a colleague upped both estimates for vertebrates to 50 percent. The US Office of Technology estimated that about one-third of exotic species end up being harmful to the ecologies they colonize. But middling odds are a poor defense against a relentless assault, nor are they a basis to be blasé about a wild card that occasionally, as in the case of zebra mussels, can literally strip the ecological gears of a river and compromise its native fauna, including its important migratory fishes.

Chapter 14

Giants of the Rivers: Gone Forever?

Bragging may not bring happiness, but no man having
caught a large fish goes home through an alley.

—*Author unknown*

Walk into the US Post Office at Hyde Park, New York, not far inland from
the Hudson River, and then look up at the high walls rimming the main gal-
lery. There you'll see a poorly known treasure—a series of 1941 New Deal
arts agency murals of the history of the Hudson, created not far from FDR's
Hyde Park residence. Standing out among them is a panel of a dramatic
scene from 1870: A sturgeon that appears to be about twelve feet long is
being hauled into a scow by three men, but the angry fish, partially wrapped
in netting, threatens to swamp the vessel. Another giant sturgeon appears
in the background, hanging at a landing where sturgeon are processed for

"A View of Hyde Park Landing"
Olin Dows, 1941

caviar and meat. The size of the fish and the drama of the scene may appear apocryphal, but that was sturgeon fishing in those days.

Some remarkably large sturgeon cruised through East Coast waters before they were decimated by the caviar craze of the late 1800s. Wood, writing of Massachusetts in 1634, described them as "all over the countrey, but best catching of them be upon the shoales of Cape Codde and in the river of Mirritnacke, where much is taken, pickled and brought for England, some of these be 12, 14 and 18 foote long." Even as late as between 1927 and 1935, nine weighing between 350 and 600 pounds were landed in Portland, Maine, from the South Channel, Georges Bank, Browns Bank, and other underwater features.

Log-size Atlantic sturgeon from the James River were essential to the survival of the Jamestown Colony, the first English settlement in the Americas, as both food and the colony's first commercial product. Archaeologists recovered thirty-two fin-spine samples from a well at the Jamestown site, plus three more from a basement. The well was dug soon after settlement in 1607 and was quickly converted to a rubbish pit, likely because of salt intrusion. Layering of other refuse allowed the researchers to determine that the sturgeon spines dated to between 1610 and 1617. When the researchers estimated the ages of the sturgeon from the growth rings in the spines, and also compared their length and width with modern sturgeon from the James, they found that the fish landed 400 years earlier averaged nineteen years and reached forty-two years, compared with eight years and reaching nineteen years in the current population—a dramatic difference. But they also found that the earlier fish grew more slowly. Why would this be? One theory was that the early 1600s were part of the Little Ice Age, which may have retarded growth. It is also conceivable that those densities of sturgeon were much higher in the then almost pristine and lightly fished river, and that there simply was higher competition for food.

American shad also are distinctly smaller today and have been for some time. Changes in their sizes and the consequent alterations in human perceptions of what constitutes "big" are especially well chronicled for the Susquehanna River. For the upper river before the Nanticoke Dam was completed in 1830, the thousands of shad taken at Rockefeller Fishery below Danville were said to weigh from three to nine pounds. Jameson Harvey, who lived near the Stewart Fishery above Nanticoke, stated in an 1881 interview: "Some of the shad used to weigh 8 or 9 pounds. I saw one weighed on a wager turning the scale at 13 pounds."

For several fisheries above Wilkes-Barre, the average weight placed at eight pounds, with a maximum of twelve. After 1850, though, lower river shad weighing more than six pounds were widely considered "large." McPhee wrote that in the 1840s, a standard pork barrel held forty shad, but that at the end of the 1860s, the same barrel held a hundred. In 1876 the *Lancaster Daily Evening Express* reported that a gentleman brought two shad to the city from below Columbia weighing six and seven pounds. The account went on to say they were very large fish, and that few of that size were ever taken in a season. By 1887 perceptions had changed considerably; the *Columbia Spy* reported on May 28, 1887, that a "shad worth seeing" was exhibited at Seneca and Mahan's fish house in that town a few days prior. Twenty-nine inches long and weighing ten pounds, it was said to be twice the weight of a large shad, or three times that of an ordinary one. By 1896 the average size of shad landed fell dramatically, to only two to less than five pounds. A modest five pounds remains the typical upper end throughout their range more than a hundred years later.

Fred Buller believed that "to catch a salmon weighing over fifty pounds on fly is the highest distinction that a salmon fisherman can achieve in his or her lifetime." This led him on a quixotic quest—to document as many Atlantic salmon of that size class by fly fishing or any means possible, resulting in *The Domesday Book of Giant Salmon*. "Domesday" or "Dooms-day" or "Judgment Day" refers to a record of English property created in 1085–86 under William the Conqueror, just twenty years after the Norman Conquest, a document that carries such weight that it is still considered admissible legal evidence today. Among the notable resources recorded in the *Domesday Book* was an abundance of salmon.

Buller's uniquely authoritative data show a sort of bell-shaped curve for truly huge salmon, those over 60 pounds, taken from anywhere within their range. Between 1700 and 1800, eight were caught. Between 1801 and 1850, five were taken, but this jumped to nine between 1851 and 1875, and to twenty-four between 1876 and 1900. The golden era was between 1901 and 1925, at fifty-four 60-pounders seen, with catches dropping to twenty-eight between 1926 and 1950, twenty-one between 1951 and 1975, and only twenty over the next quarter-century.

How to interpret this slow increase in giants, followed by a peak and then a decrease? The ramp upward likely reflects improved commercial and recreational fishing gear during times when salmon still ran in reasonable numbers. The first part of the twentieth century was where the trend lines between adequate gear and adequate numbers crossed. The lessened catches of giants in more-recent times are a sign of lowered salmon abundances and, maybe, growth potential, perhaps mitigated upward slightly by even more sophisticated approaches to fishing.

Striped bass is one anadromous fish in which size records are still creeping upward. Charles Church set out with his brother-in-law Carl Kraut in a small boat with sail and oars on an August day in 1913 to fish the coves of Nashawena Island, one among the famed-for-striped-bass-fishing Elizabeth Islands of Massachusetts. Kraut manned the oars to keep the boat perpendicular to the surf while Church cast a live eel with his bamboo rod to a cluster of boulders near shore. Immediately, a striped bass engulfed the eel, and by the swirl it made on the surface, Church knew it was big. The fish ran to deep water while Church attempted to slow it using his thumb against a pad of twine that braked the spool as the fish peeled off 100 yards of line. Then the two anglers had to maneuver quickly to avoid fouling the line as the striper rolled and headed under the boat and back toward shore. There the striper ran past a rock that could have severed the line, but Kraut rowed quickly to change the angle.

The conditions in these shallows were challenging. Church later wrote: "It was awfully rough and the boat would ride way up in the air on some of those swells so I could hardly keep my feet. I held the rod just as high as I could, and we took in about a half barrel of water, when the fish started for the boat, going out across the stern for the southeast and then off-shore for about 150 yards, for I did not attempt to hold him very hard, as I wanted to get offshore myself. I was getting tired and Carl was all wet, as well as tired. The fish lay quite still 'til I reeled almost up and down on him, and we were in seven fathoms of water, then he would run a short way, but he was getting tired like ourselves." Church tried to force the fish to the surface, straining the rod to near its breaking point, Kraut yelling, "For cat's sake, don't lose him!" The team backed down on the exhausted fish, but in the waves and current it took three tries for Kraut to finally gaff it. Church exclaimed, "Some fish!" He knew it was special because Kraut normally lifted a big striper into the boat; this one he had to roll in over the side. How special? An All-Tackle World Record seventy-three pounds; a catch that would stand for fifty-nine years.

In 1967 Charles Cinto tied Church's record with a 73-pounder trolled in the tide rips off Cuttyhunk, another of the Elizabeth Islands. But this fish did not crack the record books because it was caught with wire line and a lure with treble hooks, which is against the rules of the International Game Fish Association (IGFA). In September 1982 Al McReynolds ventured onto a jetty in a storm at Atlantic City, New Jersey. In rough surf, after a nearly two-hour battle, he somehow managed to steer the giant striper he'd hooked to the sandy shore where he beached it. It weighed 78.5 pounds. Twenty-nine years later, Greg Myerson was fishing with live eels for bait off his boat near a reef in Long Island Sound off Clinton, Connecticut. After a twenty-minute battle, he landed a striper that weighed in at 81.88 pounds, and is now recognized by the IGFA as the All-Tackle World Record.

However, a sleeper angling record may have been caught decades earlier in California, where striped bass were introduced from the East Coast in 1879. On the wall of the Marin Rod & Gun Club, a photo of a mustached man with a rod in his hand, standing near an immense hanging striper, has an old faded label on it that reads CHARLES BOND, STRIPED BASS 87.5 POUNDS, 5'7" LONG, SAN ANTONIO CREEK, CLAMS FOR BAIT.

The upward climb through the past century of angling records suggests that striped bass are an exception to anadromous fishes no longer becoming giants; larger ones have been seen and caught in nets not that long ago. The Maryland Department of Natural Resources has a mounted ninety-two pounder on its wall taken from Chesapeake Bay in the mid-1990s. And the noted writer on Chesapeake Bay, Tom Horton, saw a striper in the spring of 1992 that may have been even larger, taken in the Nanticoke River in a net on a biological survey. She was a survivor, a big fish with one eye clouded. Horton told me: "We didn't weigh it, but a couple of very experienced fisheries biologists there were pretty sure it would have gone a hundred or better. We had already pulled up some exceeding fifty pounds, and it was way bigger. I do recall its tail was about as broad as my size fourteen shoes were long." If so, she appears to have been a solitary giant. Even though the angling records have increased several times in recent decades, they are still more than a stone shy of 100 pounds, and well below the largest striped bass ever seen.

The true striped bass behemoths were encountered long ago, late in the nineteenth century, with the heaviest for which there are definite records being *several* of about 125 pounds taken in Edenton, North Carolina, in April 1891. There also are records of one weighing 112 pounds that was caught at Orleans, Massachusetts; one of slightly more than 100 pounds in Casco Bay, Maine; and one of 105 pounds at Avoca, North Carolina.

What is the future for anadromous fishes once again reaching great sizes? Poor, according to some surprising analyses, and not for the obvious reasons such as habitat degradation reducing growth potentials and overfishing causing the shortening of life spans. Recent findings show that humankind is exerting veiled but meaningful biological pressures against fish growth, both directly and indirectly.

David Conover, of Stony Brook University, built a strong reputation as a fish ecologist with highly controlled experiments with a little marine prey fish known as silversides, one of the kinds of food fish sometimes consumed as "whitebait." Conover realized that silversides could be the laboratory fruit flies of the fish world with their small size, one-year life span, and rapid reproduction. In 2006 he and his colleagues published a paper with the wonkish title "Maladaptive Changes in Multiple Traits Caused by Fishing: Impediments to Population Recovery," which reported on experiments that had enormous ramifications for the general endeavor of fishing. A catchier name for this phenomenon was later proposed: "Darwinian debt."

It is well known that continued high-intensity fishing tends to remove the individuals that grow fastest to a harvestable size. Do this for a long enough time, and this artificial selection favors slower-growing fish, shifting the genetic frequencies in their direction. That is worrisome enough, and it suggests that overfishing will lead to diminishing returns for that reason alone. But Conover and his colleagues also showed that the consequences could be far worse. They set up tanks with captive populations of silversides and then proceeded to "fish" them under different harvest regimes for four generations. In one treatment they removed the largest 90 percent in each generation; in another, the smallest 90 percent; and in the third, a random-size 90 percent.

The resultant growth was as predicted: The biomass of the large-size harvested group decreased twofold and evolved a lower growth rate, while the small-size harvested population achieved the reverse. But the critical findings occurred when they examined a suite of other characteristics of these groups in the fifth generation: Harvesting the largest individuals caused substantial declines in egg size, larval size at hatching, rates of larval growth and viability, food-consumption rates and growth efficiencies, willingness to forage under threat of predation, fecundity, and even their average number of vertebrae. This is contrary to classical fisheries thinking, which says that harvest-reduced reductions in population size will increase available resources for the survivors, thereby increasing their fitness. This is the crux of the argument that mortality from power plants induces compensatory effects in the remaining population. But a simple experiment on a little fish is indicating that no, fish subjected to long-term harvest have a Darwinian debt to pay. Why did fishing's artificial selection on size cause such mayhem on these critical survival characteristics of the silversides? What was missed till then was that many of these traits are linked genetically, so that when large size underwent "Darwinian selection," a host of correlated traits went along for the ride. The diminutive silversides of generation five revealed that a fished population may be forced to sustain itself having suffered considerable damage to the fundamental characteristics that allowed it to flourish prior to fishing.

Fishing can impose immense evolutionary "force"; fishing mortality may exceed natural mortality by 400 percent. When applied long enough to skew the demographics of a population, the mere cessation of fishing doesn't cause an immediate reversion to the earlier conditions; this "debt" of genetic change must be paid over many generations as genotypic frequencies come back to a normal balance. Many important fisheries, including for river-sea migrants, have failed to bounce back even when fishing has been drastically curtailed. Part of the answer may be this cryptic but possibly critically important fisheries-induced evolution.

If the effects of fishing aren't enough, climate change may help to spell the end of outsize individuals; indeed, if the forecasts by a study published in the journal *Nature Climate Change* by William Cheung, Daniel Pauly, and a half-dozen other coauthors are accurate, the ability of the oceans to support fish populations at physical body sizes we are accustomed to by 2050 is going to be curtailed. And, though the scientists did not speculate on this, these size reductions may have profound effects on characteristics such as egg production that may reduce abundances and, therefore, fisheries yields.

In this ambitious study, the researchers examined the effects of warming seas on the maximum predicted size of some 600 species of fish from the world over. They did this using a not-unreasonable maximum-emissions forecast of warming in conjunction with a well-known formula that relates fish growth to available oxygen. Because of the fact that the warmer the water, the less oxygen it can hold, higher-temperature seas would be expected to slow growth down and, indeed, this is what was found. But what was both unexpected and worrisome was the sheer *magnitude* of the forecasts. Most broadly, for fish species assemblages in given regions, the average maximum body weights can be expected to shrink by 14 to 24 percent. About half of this is due to changes in distributions and abundances, but the other half is due to their physiologies being compromised by lowered oxygen levels. These predictions are not the same across all latitudes; tropical and temperate regions will see the greatest effects, of more than 20 percent, but the phenomenon will be felt globally. If their calculations are correct, the giants of the rivers will be gone forever.

Chapter 15

Peering into the Black Box

Follow the river and you will find the sea.
—*French proverb*

A "black box" is a concept in science for a link in a system where what goes in the box is known and what comes out is known, but what happens *in the box* is a mystery. The black box is an apt metaphor for what diadromous fish do at sea.

It's late on a comfortable early May evening as I follow Steve Gephard's directions to an iron gate near the Rainbow Dam of the Farmington River. The Farmington is a major Connecticut River tributary that enters the big river near Hartford. At one time the Farmington supported great runs of Atlantic salmon, shad, and other anadromous fishes. Today it is one of a network of Connecticut River tributaries in which multitudinous salmon fry are stocked in the hopes that they will generate a respectable spawning run from a sorry baseline. From Connecticut being a productive salmon river in the 1700s, by 1872, when a solitary salmon strayed into a fisherman's net inside the mouth of the river, no one could identify it. Hope is one of the essential ingredients that help to sustain these modern efforts, but almost fifty years of trying have yielded strikingly poor results.

A fish ladder that tops the dam boils with water crashing through it. This is no small ascent for a fish to make—even I'm huffing as I climb the fifty-nine human steps that parallel the sixty-vertical-foot-high fish ladder. But this evening I am more interested in a downstream migration, the mass movement of young salmon that have survived and grown since being stocked as larvae and that are now responding to the suddenly urgent physiological imperative of smoltification. Barely more than a half-foot long, the smolts have turned silver in preparation for life in the sea, and they are primed to adapt to rapid salinity increases from the sweet water of the upper Farmington to the full-strength seawater where they will mature—if they endure. When I arrived at 8:15 p.m., though, there weren't any outmigrating salmon in the holding pool where they are kept until they can be processed for the salmon-recovery monitoring program.

The young salmon are reared over much of the Farmington watershed where they mix with the trout that are pursued by fly fishermen in its bucolic upriver reaches, but as their instincts instruct them to migrate downriver, they enter the quiet waters of Rainbow Reservoir. Two state biologists, Tim Wildman and Amy Koske, are ready in a work shed to count, measure, and tag any salmon smolts that manage to find the exit chute near the top of the dam and ride the flume down to the holding pool. What a difference a half-hour and a tad less light make. When I come back down the steps at 8:45 p.m., just as full darkness is taking hold, the pool is alive with frisky young salmon, with more entering every minute—a mass nocturnal migration to the sea.

It's time to process fish. Wildman jumps into the holding pool with a square-panel net and herds some of the fish into the work area that includes a trough with flowing water, and then he and Koske begin to work up the catch. The salmon are magnificent examples of their kind, nicely proportioned, well fed, healthy-looking, and I wonder: Why isn't this Connecticut River salmon-restoration program working? During this time of year, the Farmington and a number of sister tributaries from here to Vermont yield their treasure of young salmon to this seaward migration, as many as 1,800 smolts in one night passing over the Rainbow Dam, and yet the total number of adult returns each year to the entire Connecticut River watershed hovers around 50. Clearly, the problem is not in this stage of life, which leaves only the great black box, the marine phase. Wildman makes a wry joke about this, saying that "Salmon do okay in Connecticut fresh waters; maybe we should keep them from the sea."

About an hour after dark, we get a surprise: There is something really big swimming with the smolts. Wildman jumps in and fishes it out—a starved-looking and deformed adult salmon, with white fungal patches on its snout. Likely a leftover broodstock fish from the salmon-hatchery operations stocked in an inland waterway to offer excitement to freshwater anglers—a fish so awful-looking that mercy killing seems to be in order.

The Connecticut River Migratory Fish Restoration Program is about as massive and complex a fish-restoration government program as they come. Five state and four federal agencies are involved. Baby salmon stocked in the watershed can originate from hatcheries at Roxbury, Bethel, and Pittsford, Vermont; Warren, New Hampshire; Hartsville and Palmer, Massachusetts; or Riverton and Kensington, Connecticut. Yet this gargantuan effort results in only those fifty or so adults returning. What happens to salmon at sea?

Atlantic salmon smolts embark from the relative safety of the Connecticut and hundreds of other rivers across the arc of lands rimming the North Atlantic Ocean and forage and grow in its dark, unsettled waters. The salmon appear at certain oceanic locations at times well known to commercial fishermen, with intensive netting for them beginning in 1960. There once were well more than a thousand salmon rivers spread across the United States, Canada, much of Western Europe, and on great islands such as Greenland and Iceland. When salmon were still abundant, the particulars of where the salmon actually emanated from didn't matter that much; there was no shortage of fish. In North America alone, the historical levels of spawning-size salmon were 2.5 million to 5 million. But a seemingly unstoppable spiral brought their numbers by the mid-1970s down to about 800,000. And the crash continued: 300,000 by 1991, 125,000 by 1996, and fewer than 70,000 in 1999. As catches plummeted, the salmon managers of each nation wanted to know whether it was their fish being taken in these marine waters, because each salmon landed at sea was one fewer that could return to their rivers. And declining salmon meant declining commercial catches in rivers and angry fishermen, and declining sport catches in rivers and angry fishermen.

Many scientists and managers believed the problem causing the plunge in salmon was occurring in the marine phase of its life cycle, again, the black box problem. But how to study a relatively scarce, high-level predator that roams across endless miles of unfriendly seas? Michael Dadswell became intrigued by the problem. Dadswell is an iconoclastic professor at Nova Scotia's Acadia University who relishes thorny fish-related problems and issues, even if they lead to acrimonious debate. When the Tidal Power Corporation proposed harnessing the potent tidal forces of the Bay of Fundy for electric power generation in the early 1980s, Dadswell led the grassroots pushback which pointed out that bay-wide barriers to harness the energy from tidal flows would cause dire harm to the various anadromous fishes that use the Bay in summer as part of their migratory cycles, and that this would be too high a price to pay for more power. Dadswell had an abiding fascination for the mysterious plunge in salmon numbers. He would talk knowledgeably about all the hypothesized causes of salmon declines and make a case against each of them. What was left was illegal high-seas or "pirate" fishing, something hard to document in the vastness of the North Atlantic.

Dadswell also wanted to understand the pattern of the salmon's marine movements, to go beyond simple descriptions of almost random dispersion to one linked to his intuition: that the physical and biological realms were tightly linked for the species, the product of long-term evolutionary inter-play in which salmon, like many other fish, exploit the physics of the system. One hypothesis was a model in which salmon essentially swam to the various edges of the central North Atlantic feeding grounds from their home rivers and then swam back again. Enough tagging data had accrued through the years which supported at least the possibility that salmon migration was that simple. The second model, which Dadswell tested, was the "Merry-Go-Round Hypothesis," proposed in the 1980s, in which salmon make trans-Atlantic migrations using surface currents of the North Atlantic subpolar gyre. In this visualization North American and European salmon enter the gyre from their respective shores and migrate counterclockwise around the North Atlantic, feeding, growing, and mixing until they mature, and then use the gyre to return to their home rivers. Dadswell's and his col-leagues' review of many bits of ancillary information from a medley of catch data and scientific studies came down squarely for the Merry-Go-Round Hypothesis—one which implies that salmon from most rivers rimming the North Atlantic are vulnerable to commercial harvest at many high seas locations.

With salmon from hundreds of rivers swimming around in and perhaps cycling through the Atlantic, in the latter part of the twentieth century, commercial fishermen positioned on the hypothesized merry-go-round helped world landings reach as much as 25 million pounds per year, a toll that then showed up as ever-declining runs in the fish's home rivers. One man, Orri Vigfússon, was moved to try an innovative approach. Vigfússon had the perfect confluence of experience and interests to pull it off as a native Icelander, an entrepreneur, a lifelong outdoorsman who "fell completely in love" with salmon fishing, and someone who in the 1970s became deeply concerned about the shrinking salmon runs in the northern Icelandic rivers he fished. His ambitious solution? Vigfússon launched a program in 1989 to buy out the fishing rights of commercial salmon fishermen whose overfishing was causing the decline. The North Atlantic Salmon Fund (NASF) has since raised US $35 million to buy the netting rights from commercial fishermen across the North Atlantic, paying them not to fish salmon for a set number of years. The expectation is that over time, fewer and fewer fishermen will wish to return to netting salmon. To promote their leaving the fishery permanently, a large portion of the funds of NASF are devoted to helping fishermen find alternate employment.

The NASF program was a radical step and positive results did not come easily. Vigfússon appealed to stakeholders across Iceland, Europe, and North America to convince them of the need to strongly address the overfishing problem. He met with anglers and residents of river communities, who were all experiencing declining numbers of river-run salmon. He reached out to commercial salmon fishermen, talking openly about the extent of the problem from both an environmental and economic point of view, including how their own livelihoods were being affected. After raising significant grassroots support, Vigfússon approached governments, introducing his idea of the buyout agreements.

Given the high value of wild salmon, resistance was strong. But with a mind for business and intense passion for his cause, Vigfússon brokered multimillion-dollar buyouts or moratorium agreements with commercial salmon fishermen in the Faroe Islands, Iceland, Greenland, Wales, Ireland, Scotland, England, France, and Norway. The NASF estimates that commercial fishing in the open Atlantic has dropped by more than 75 percent in the last fifteen years, and river anglers in several countries in areas where nets have been closed have reported substantial increases in salmon catches. More than 5 million North Atlantic salmon have been saved to date, according to NASF.

Vigfússon says: "The life cycle of a salmon is something spectacular," and that "My mission in life is to get back the salmon stocks into historic abundance." He's doing this by delving into the black box. Deservedly, Orri Vigfússon has been called the "most honored angler on Earth."

Despite Vigfússon's extraordinary successes on the high seas, the relict salmon populations of Maine also face a recent existential challenge emanating from coastal waters: escapees of salmon from aquaculture facilities—now rated by the Maine Atlantic Salmon Task Force as the second greatest threat behind dams. Virtually all of the more than 1 million tons of Atlantic salmon being sold in markets worldwide is farmed in pens set in estuaries and bays from Alaska to Washington to Chile to Norway to Scotland to New Zealand. Some also are ranched within the natural range of Atlantic salmon in North America. Why would having farmed salmon positioned near wild salmon pose a threat? Remember that the word *salar* in its name means "the leaper," and leap they do, often escaping their pens. Holding nets may also be blown apart during severe weather. Once they make it into the wild, these domesticated salmon—mainly originating from a Norwegian strain and bred for fast growth, not for the myriad naturally selected characteristics of wild fish—often migrate into rivers where they compete with or breed and genetically degrade the native salmon.

Large escapes have occurred off Nova Scotia near the small Downeast rivers of Maine where relict stocks of salmon hang on. Indeed, in 2001, 62 salmon of aquaculture origin were captured in Maine's Dennys River—this, in a system with a historical run size of only 150 to 450 salmon. Compare those numbers with the scale of farming; some pens may hold as many as 90,000 fish. Massive escapes have occurred elsewhere; in 2005 alone, some 600,000 off Scotland and nearly a half-million in Norway, these often storm-related events contributing to an estimated 3 million escapes annually worldwide. Scientists estimate that as much as 90 percent of the salmon now found in some Atlantic rivers are fugitives from ranching operations.

Ranched salmon, living literally penned in tight quarters, are also vulnerable to epidemic diseases, which they may spread to wild salmon through the mesh of their enclosures, or as escapees that mingle with them. For instance, fugitives carrying the infectious salmon anemia virus (ISAv) threaten wild fish. Salmon with ISAv succumb at a high rate; treatment requires total eradication of the salmon stock on any farm infected. Farmed salmon also spread sea lice. Sea lice are unappealing-looking invertebrate parasites that weaken salmon; even two or three can kill a juvenile fish. Though found on wild salmon, they rarely are abundant enough to cause harm. But salmon ranches can support major sea lice eruptions that then infect and harm wild salmon stocks. Several decades of experience suggest that taking the harvest pressure off the wild salmon through Atlantic salmon ranching has not done the wild fish any favors.

Once, the Gulf of Maine's inshore waters were home to untold numbers of codfish. All evidence indicates they were divided into numerous local races associated with particular reefs and other bottom structures, and that they had various body form and behavioral differences the commercial fishermen who harvested them were familiar with. Most of these stocks crashed and essentially disappeared by the mid-1900s.

A unique aspect of some of these cod stocks was that a portion of them were intimately tied to the alewife migrations. But populations of both species were reduced to relict sizes or extirpated within the past thirty years. To determine whether these declines were related, a team of historical ecology researchers from Stony Brook University delved into records and other data sources for the Gulf of Maine back to the 1600s. Accounts from 1776 and 1824 link the choking off of alewife runs from milldam construction to declines of cod, haddock, and other sea fishes, the latter stating that "it is well known that the Cod follow the alewives in great numbers, even into the Bays and Harbours where they frequent." Spencer Baird, later the well-respected US Commissioner of Fisheries, after interviewing fishermen in 1874 wrote that he was inclined to believe "that the reduction in cod and other fisheries, so as to become practically a failure, is due to the decrease off our coast in the quantity, primarily, of alewives . . . more than to any other cause." He stated what seems surprising now but which he then took for the obvious: "We are all very well aware that fifty or more years ago, the streams and rivers of New England emptying into the ocean were crowded, and almost blockaded at certain seasons, by the numbers of shad, salmon and alewives seeking to ascend, for the purpose of depositing their spawn, and that, even after these parent fish had returned to the ocean, their progeny swarmed to an almost inconceivable extent in the same localities, and later in the year descended to the sea in immense schools." He added that cod and other groundfish could then be caught from small open boats only a short distance from the fishermen's abodes, and that at the time of his writing, anadromous fishes were "almost unknown in many otherwise favorable localities."

The Stony Brook team did find a statistical association in the late twentieth century between alewife abundance and cod recruitment. Who would think that a deepwater marine fish like the cod and an upper-water-column, anadromous fish like alewife would be intertwined? But this intimate connection is just a more direct association than the many other myriad ways in which these two aquatic realms are dependent on each other.

The July 31, 1856, edition of *Country Gentleman* contains a lighthearted but informative article about eel fishing in Moriches Bay, one of the brackish water bodies behind Long Island's barrier beaches. The gear is described, as is the recommended bait: horseshoe crabs, quartered. Only forty traps kept a fisherman busy, with two trap runs per day, with 25 dozen eels the usual daily haul. The catch was sold to a middleman, who had the eels skinned. The skill of the skinners was extraordinary. The eels were kept on sand to give the skinner a grip on their slimy epidermis. A quick slash of the throat all the way to the neck skin, followed by a belly slit, and then with the thumbnails, a quick "artistic" jerk and the eel skin was off. Remarkably, a good skinner could strip the skin off 30 dozen in an hour. The daily shipment by railroad from Moriches Bay to New York City in those days was 300 dozen, and this kind of fishery went on all over Long Island.

In 1938 a biologist, J. R. Greeley, with the New York State Department of Environmental Conservation supervised a survey of the inshore coastal waters of Long Island. Although they used a seine net, not really the best eel gear, all summer long they captured two or three eels per short haul. Eels clearly were still everywhere, in large numbers. In 2010 my master's student, Peter Malaty, repeated Greeley's Long Island survey, working the same locations on approximately the same dates with similar gear. Malaty and his crew caught exactly one eel the entire season.

There is anecdotal evidence that variations in ocean currents and circulation patterns across decades, perhaps exacerbated by climate change, are affecting movements of young eels into fresh waters. A climatic phenomenon, the North Atlantic Oscillation (NAO), is an indicator of regional conditions in the North Atlantic in the same way the more famous El Niño Southern Oscillation is for the Pacific. A correlation between the variation in the NAO—the difference between sea-level atmospheric pressure at the Azores or Lisbon, Portugal, and at Reykjavik, Iceland—and reduced eel recruitment has been found both at Den Oever in the Netherlands and at the Moses-Saunders eel ladder in the St. Lawrence River, continental bookends that suggest that as NAO values increase, eel recruitment decreases. High NAO values are associated with periods of above-average winds, increased mixing of surface water, and easterly shifts, combined with a reduction in speed of the Gulf Stream and North Atlantic Current—the eel elver delivery systems of the North Atlantic Ocean. The influences of the NAO on larval eel delivery may be most pronounced at the extremities of the geographic range, such as within the St. Lawrence River and Mississippi River watersheds.

Oceanic currents may also be increasingly influenced by global climate changes. Larger freshets of melted freshwater ice may slow or divert the Gulf Stream current farther offshore and negatively impact eel transportation.

Reduced survival from a slowed or diverted current system may result from starvation, increased predation, or a breakdown in the signal and timing between glass eel transformation and detrainment from the Gulf Stream toward the coast.

The black box that is the ocean environment is called that for a reason. But recently society has shone brighter and brighter lights into its void, and it hasn't been pleased with what's been revealed. The oceans are changing physically in well-documented ways that will perturb their biology, namely acidification, sea-level rise, nitrogen overenrichment from rivers producing "dead zones," and temperature shifts due to climate change. A flurry of studies has shown that marine life is going to be altered and compromised in numerous ways. This will include not only those innumerable obligatory sea creatures, but also that select and special group of migratory fishes that disappear into the black box as they enter the more mysterious facet of their grand life cycles.

Chapter 16

Hatchery Stocking: Subtraction by Addition

We cannot turn away to contemplate
The normal; we must hover still and stare
At what is so easy to create
And just as easy to eliminate—
Eager translucent bodies, unaware.
—*Celeste Turner Wright, from "In the Fish Hatchery"*

It's a warm late-May evening and that rich, pleasing, big-river smell fills my nostrils as I wait for the shad crew to arrive. Indeed, the Mighty Delaware River seems uber organic with splashing fish and all kinds of bird and bug life as it slides past me with quiet strength at one or two miles per hour, just fast enough to roil its surface. Though I am at the Smithfield Beach boat-launch ramp in Pennsylvania, the other shore is New Jersey, but it's not a stereotypical suburban "Jersey" scene. Before me lays the still-wild Tocks Island Pool, a major spawning reach for American shad, and the former battleground for a major damming controversy.

Soon appearing are two pickup trucks filled with men from the Pennsylvania Fish & Boat Commission, pulling two boats loaded with all kinds of gear. There is no rush; the crew, led by Daryl Pierce, knows from long experience that catching shad here is a night game. Being a little early and fish guys at heart, they can't resist taking a few casts with rod and reel before they break out the serious equipment.

Bryan Chikotas, a no-nonsense crew member, shows me how the 200-foot-long gill nets will be set by drawing lines with his index finger in the mud. Not perpendicular to the current as in commercial operations seeking to make the largest catches, but instead, in several overlapping rows parallel with the shoreline, little more than a boat's width apart, and not far upstream from Tocks Island. With this scheme some small pods of rushing and courting, spawning-obsessed shad will dart toward the suspended nets and become entangled. Unlike so many rivers from Maine to Florida, shad are abundant here. The undammed main-stem Delaware is one of the few bright spots among the species' many declining populations. As the crews motor off from shore to begin work, V-wakes are already visible here and there as shad ramp up their procreative rituals. In the darkness their spawning can be heard as splashes on the river's surface as males chase females ready to spawn.

I am assigned to net-picking duty. We slowly maneuver along the outer edge of the net illuminated by battery-powered spotlights, following the line of donut-shaped buoys to look for shad caught in the netting. The crew members jokingly point out that I don't yet have an eye for this as they spot fish I've failed to notice. They can read the height and curvature of the buoy line ahead to tell where the next shad is in the gill net, and unlike me, they can untangle it from the mesh in a second or two. Despite being a marginally useful greenhorn, though, I am elated at the chance to help gather the silver bounty of the Delaware.

I'm also deeply moved at our collecting the shad in this very location in the Delaware Water Gap, for we are on the famously contested Tocks Island Pool. If fate had turned out differently, this bucolic stretch of river would have been submerged under a thirty-seven-mile-long reservoir. The Tocks Island Dam had slumbered on the books of the US Army Corps of Engineers as a possibility for the Delaware in the great dam-building era, but the immense floods and one hundred deaths from Hurricane Diane, which followed rising waters from Hurricane Connie by only one week, provided strong momentum for a proposal to Congress in 1965 for a multi-purpose dam that would have provided hydroelectric power, drinking water for New York and Philadelphia, a National Recreation Area, and flood control. In preparation for the dam and the giant reservoir behind it, the federal government began to acquire valley-bottom land, sometimes through condemnation. This was the time of the flowering of the environmental movement, however, and just as on the Hudson River, local citizens began to fight the taking of their properties for a project that would end their way of life and cripple their river. The larger question being asked, ironically: "Was the river to serve mankind, or remain free-flowing?" As if a free-flowing river didn't serve mankind in myriad ways.

The battle against the dam was pitched, and included the strident support of Supreme Court Justice William O. Douglas, who fell in love with the area during a visit, and determined hippie squatters in a condemned village that stayed for several years, even after some were forcefully evicted en masse one night in 1971 by federal marshals and bulldozers that razed their homes. Though these skirmishes were colorful, the real foot-draggers were sequential governors of New Jersey, William Cahill and Brendan Byrne, both of whom saw issues concerning land acquisitions and environmental problems, and possibly better alternatives for flood control and water supply. In 1975 three of the four states represented in the highly political Delaware River Basin Commission voted down the project, a decision assisted by its high costs in the face of the financially costly Vietnam War, and by a bedrock base deemed too unstable to support the largest dam east of the Mississippi. And so the Delaware remains the last major free-flowing American river east of the Mississippi and, consequently, home to one of the healthier shad stocks.

Before long we have our boat resting in the shallows near the ramp, our haul filling tubs, and it's time to "make" more shad. I'm taught how to strip eggs from the freshly dead females, squeezing the fish's abdomen between my thumb and index finger from behind the gills down to the vent till the golden stream yields only a trickle, perhaps 30,000 eggs extracted in seconds, and then on to the next fish. Then sperm is added and the mixing and marriage of the two gently encouraged using a feather. Particles of blood, now clotted, that help nourish the eggs are picked out to avoid spoiling the batch. It seems almost miraculous that inside this milky bowl, hundreds of thousands of shad will begin their lives, part of the program's annual target of 13 million fertilized eggs.

I begin my long drive back to New York near midnight, with the team still at it. The goal this night is twenty liters of eggs, and with the spawning shad on the move, it looks like they'll make it. Later the eggs will travel to the Van Dyke Hatchery, with the young shad being used to help restore the species in two main tributaries to the Delaware, the Lehigh and the Schuylkill, the latter more than two centuries after the migrating shad of spring 1778 either saved or didn't save Washington's troops.

Hatcheries also did or didn't save America's anadromous fisheries, but mostly didn't. Why aren't hatcheries the panacea they appear to be at face value? The life of a hatchery-reared fish is a strange one. It emerges from an egg among many of its kind, and at first it is coddled, albeit in tight quarters, with full protection from predators and with regular feedings of nutritious but highly artificial food. At some point in its development—from larvae, to fingerling, to older juvenile, to yearling—it is released into the wild with its cohort. Although the site for this rude shock may be chosen so as to minimize predation and to maximize natural food availability, the "stocker" is completely naive to the wild, with only its perhaps dumbed-down instincts to rely on. It may indeed fear and flee a larger fish that would like to eat it, and it may itself pounce on a prey item, but it will do both with anywhere from a little to a lot less effectiveness than its wild brethren already steeled by nature.

In most instances the ranks of stocked fish will fall like the front lines at Gettysburg. Any planned stocking also always leaves those in charge with a conundrum concerning the fish's survival in the wild. Stock them early, perhaps as larvae, and the total numbers placed in a river are high (before attrition in the hatchery), the fish have had only a short time to become "domesticated" by hatchery life, and costs for rearing are minimized. But the rate of survival may be minuscule given their bite size. Raise them for months or a year or more, and they are larger and better able to endure, but their numbers will have diminished, they will have become more habituated to the food and safety of the hatchery, and it will have been expensive to maintain them. Somewhere the trend lines of the pro and cons cross at a most optimal level, but they remain only facsimiles of wild fish.

Once placed in the wild, the stockers may behave differently than the wild-born over extended periods of time; some anadromous fish may even "residualize," remaining as nonmigrants in rivers where they were stocked. Indeed, when juvenile Atlantic sturgeon were stocked in the Hudson River in 1994, they grew more slowly than naturally produced sturgeon and were less likely to migrate to sea, behaving like the much more riverine shortnose sturgeon.

There are even differences in condition and morphology. A good trout angler prizes wild trout partly because they fight noticeably harder. Wild trout also look better, having classically cleaner lines and few abnormalities, whereas hatchery-produced specimens look less "right," and often have a short or twisted fin or a misshapen gill cover. Hatchery fish may compete against wild members of their own species, lowering the fitness of the wild fish while not compensating fully with their own reproduction. There also is strong experimental evidence for a "domestication" effect among hatchery-reared fish. Using reproductive success in the wild as a measure of fitness, researchers found a more than one-third decline for every captive-reared generation when more than one generation is reared in hatcheries.

Formation of the US Fish Commission was linked to the notion that habitat degradation and overfishing could be ameliorated by "manufacturing fish." Stock enhancement would cure these ills through regular additions of cultured fish to wild populations. In 1872 Congress appropriated $15,000 to fund the US Fish Commission to investigate the causes of decline in our fisheries and to come up with solutions. An obvious answer was to build many hatcheries and propagate and stock billions of baby fish of many species and scatter them over the waters.

The unbridled optimism of fish culturists in their belief that they would make our waters teem with fishes is epitomized in an address made by Robert Barnwell Roosevelt at the annual meeting of the American Fish Culturists' Association in 1876. In this centennial address Roosevelt recognized the great deterioration of our fisheries but concluded that a "new science was being born into the world . . . but the clear light is visible at last. . . . There need be no fear for the future, and in much less than a hundred years, the waters of America will teem with food for the poor and hungry, which all may come and take." This followed by one year a statement by Spencer Baird, the US Fish Commissioner, who advised the commercial fishing industry that artificial propagation of salmon would be so successful it would eliminate the need to regulate harvest.

The early promise of fish culture was seen as an almost noble enterprise, one that would mitigate the effects of overfishing and allow the spread of desirable species to new waters. To achieve these goals the level of the efforts needed to be grand. George Brown Goode argued that "[p]ublic fish culture is only useful when conducted upon a gigantic scale—its statistical tables must be footed up in the hundreds of millions." That nonnative species might be harmful was as yet itself a foreign notion. To disperse the eggs, larvae, and fingerlings of the hatchery troughs and culture jars, special railroad cars were fitted to carry them to distant waters. It was a free-for-all, practiced without much thought as species were "tried" in new environments, with most failing. Smelt were released into Chesapeake Bay tributaries, in waters too warm for smelt. Lobsters and tautog were dumped into the Pacific but didn't take. Also to no avail, lake trout and whitefish from the Great Lakes were let go on Long Island, New York; European dwarf whitefish were introduced to the Penobscot; and Pacific salmon were stocked in Hudson River tributaries and the Susquehanna.

Some stocking programs worked well enough, but only as long as fish were continually dumped into these locations where they weren't naturally suited. Though there has been a long debate about whether Atlantic salmon were native to the Hudson River, this notion is likely based on a misconception by the explorer Henry Hudson's mate, Robert Juet. As the *Halfmoon* journeyed upriver in September 1609, he wrote that they saw "many Salmons and Mullets and Rays very great" and "great stores of salmons in the river."

Planking shad at Marshall Hall, Maryland, on the Potomac
William Cruikshank, 1893

No ichthyologist, I surmise that sailor Juet saw the salmon-like weakfish. But the dream to "restore" Atlantic salmon to the Hudson persisted, and several million salmon young reared from Penobscot River salmon eggs were stocked in the Hudson in the 1880s. Enough of them returned from their journeys at sea to support a small commercial fishery, with more than 300 as large as 38 pounds landed one year, but the river offered no good spawning habitat, salmon eggs became scarce and more expensive, and the experiment failed.

The newly constructed Conowingo Dam kept one of the continent's largest and longest American shad migrations from being barely able to surpass the mouth of the Susquehanna River, thereby dooming it unless something else was done to replace all that vanished reproduction. A rescue by hatcheries was the easy answer, an apparently foolproof technological antidote—the addition of manufactured fish to replace those lost to over-fishing and other issues.

Though the Susquehanna fisheries had remained strong in the 1870s, the potential of catching even more fish through fish propagation inspired industrial-scale shad production in the upper Chesapeake Bay at Havre de Grace, Spesutie Narrows, and Shad Battery Island. Within three years of commencing operations, the Maryland Commissioners of Fisheries claimed that the hatcheries were responsible for increased landings, even though few fish from those releases would have been expected to return to the Bay so quickly. From this early hint at success, the Commissioners bragged that "they are now confident of our ability to render our waters as productive of the most useful varieties of food fishes as they have ever been."

The propagation operation at Shad Battery Island soon grew to include the purchase of copious amounts of roe from commercial gillnetters, the capture of more roe via netting by the hatchery personnel, and even more through the temporary hiring of forty men and boys to pull seines and work gill nets. The yield from all these efforts in eggs was indeed "footed up in the hundreds of millions": In 1898 alone, 164,655,000 shad eggs were successfully propagated. Even though much of these plantings made little biological sense, eggs and young shad from the station were stocked in the Susquehanna and other Maryland rivers, Delaware, New Jersey, the Gulf and Pacific Coasts, tributaries to Great Salt Lake, and various inland Western lakes and rivers. Artificial propagation of the American shad in the United States peaked in 1900 at 36,749,000 eggs, 202,307,000 fry, and 2,000,000 fingerlings.

But the promise of shad hatcheries soon evaporated. Intense fishing in the head of Chesapeake Bay and farther up the Susquehanna, with no "lift days" enacted to allow shad to reach spawning grounds, and the building of additional dams shrank the runs despite the massive stockings. So rapacious were the fishermen that one report from Pennsylvania stated: "There was not merely a lack of cooperation, but a spirit of absolute antagonism, which, in a number of well-authenticated cases, manifested itself in the form of open and violent resistance to officers of the law." One of these included flipping five wardens out of their skiff. When they swam to shore they were met with "attention from the bystanders . . . in the form of blackjacks, empty bottles, and sections of rock." Soon changes in commercial fishing practices and a drop in the numbers of migrating shad made it impossible to obtain enough eggs to justify the operation of the Shad Battery Island facility, thereby negating any possible benefit it had provided, and further reducing the runs. As Seth Green and a colleague put it in 1879, "No skill in fish culture can hatch fish when there are no parents from whom to obtain eggs. That proposition is self-evident."

Nonetheless, the federal government ran the Shad Battery Island hatchery until 1917, when it blamed the State of Maryland for failing to control its fisheries, but also offering to step back in if the state would do so. Instead, Maryland simply let the fisheries continue and opened its own hatchery on a barge moored near Shad Battery Island. Then the Conowingo Dam was built in 1928—without a fishway.

By the 1970s shad runs to the Conowingo Dam were minuscule, with fewer than 300 trapped for restoration efforts per season. But soon recovery efforts were ramped up. First, a total of 211 million eggs were collected from the Susquehanna, Potomac, and Connecticut Rivers, and even the Columbia on the West Coast, where, ironically, shad had been introduced from the East Coast in 1871. From 1981 to 1986, 1,200 to 1,600 ripe adult shad were trucked from the Hudson River each spring and stocked to spawn above Susquehanna River dams. By 1986 shad numbers below the Conowingo had increased sufficiently that all subsequent transfers by truck upriver were trapped there. From 1985 to 2000, more than 166 million shad larvae and fingerlings also were stocked from six East Coast rivers and the Columbia in a comparative "performance evaluation." (The tough New Yorkers from the Hudson were the clear winners.)

With all the stocking efforts generating more shad in the lower Susquehanna, a deal was reached in 1988 between the Conowingo Dam operator and resource agencies to build a fish lift on the face of the dam. Optimism reigned; the elevator was commodious, ready to handle 750,000 shad and 5 million river herring. Fish-passage facilities followed soon after at the next three dams upriver, designed to get a half-million shad farther toward their primordial spawning grounds. Only a meager goal compared with original runs, but orders of magnitude higher than what were actually occurring at that time.

This highly technical wedding of stocking and lifting worked for a while. Between 1997 and 2000, 354,133 shad passed Conowingo, and more than 82,000 made it past the fourth dam. Chemical marking of the fish's ear bones before propagated shad were released to the wild showed that in the late 1980s and early 1990s, shad born in hatcheries made up 80 to 95 percent of the run, but that not long after, fish spawned in the wild river grew to as much as 90 percent of the total. It was not destined to last, however; the number of shad taking a ride in the hopper from the river below the Conowingo Dam to the reservoir above declined from more than 100,000 in 2004 to little more than 22,000 in 2012. And only 224 made it past the fourth dam, York Haven, although this was somewhat better than in 2011, when no fish passed it.

Some hatchery effects are just plain weird. In 1991 I collected blood samples for DNA analysis from striped bass from the ocean surf on eastern Long Island, New York, in a program being run with commercial fishermen and overseen by Victor Vecchio, a biologist with the New York State Department of Environmental Conservation. My mornings with this crew were among the most memorable of my fish-collecting career. Vecchio and his agency were running a "fishery-independent" striped bass stock-assessment program for the Atlantic States Marine Fisheries Commission, to take the pulse of ocean-dwelling striped bass from the migratory populations, mixed together, at a strategic location they passed on their annual autumn migrations southward. *Fishery-independent* meant the use of a gear type that was less likely to show gear-type influences on the resultant catches—that is, size and age selectivity—thus offering a realistic "snapshot" of the demographic makeup of the total migratory striped bass stock. The technique was haul seining, the same approach, but with major variations, that had been used along the Eastern Seaboard since colonial times.

It was the combination of the fishing technique, the habitat, and the fishermen that I loved, the withering end of an era and a colorful way of life. We'd meet for coffee at dawn and then caravan along the sandy ocean beach at Montauk or Amagansett while we studied the surf to determine if we should brave it. The fishermen were "Bonackers," the local name for the crusty, independent baymen who spoke with vaguely Elizabethan accents and whose ancestors had lived in the region near Accabonac Harbor since the 1600s. Their haul seining had been controversial, a "surf and turf" issue between shore fishermen they referred to as "sports" and the Bonackers, who sometimes brazenly set their nets around schools of stripers and blue-fish the sports were catching. The baymen were often their own enemies by not carefully releasing all the "trash fish" that were landed among their commercial species. This allowed sportsmen's groups to demonize the prac-tice of haul seining as a "dirty" fishery. In truth, a haul seine simply herds fish to shore and, if done carefully, leads to little mortality. With battle lines drawn and a declining migratory striped bass stock in the 1980s, haul sein-ing on Long Island was banned. Except not quite; haul seining was dying a slow death because New York State hired one crew each autumn to conduct the fishery-independent survey.

As we drove the beach, if the waves were of the right height and were found to be breaking nicely on a certain stretch, the baymen would posi-tion a pickup truck and a double-ended dory on a trailer perpendicular to the strand, rev up the outboard motor that rode in an engine well inside the dory, so as not to take a wave and cut out dangerously, and, when the decision-maker in the boat saw a break in the waves he liked and gave the signal, the truck driver would step on the gas and back down at high speed to the edge of the sea and stop short. With this momentum the dory would leap off the trailer's rollers, the motorman in the dory would gun the engine and charge through the waves, and, moments later, hundreds of yards of netting would be fed into the sea in a broad arc. But the adventure wasn't over yet. The dory would be circled back to shore downcurrent, and then a path would be picked through the waves and the dory beached, before being pulled back up on the trailer. Then the real work began: retrieving an enormous net against tide and longshore flow and, if fortunate, filled with hundreds or thousands of pounds of fish.

As awe inspiring as the setting of the net was, this next stage was pure artistry. Lines at each end of the net were run to two trucks that were spaced wide apart, roughly the width of the set. Long ago the Bonackers learned that the great strain from the net would whip an electric winch in no time, so they ingeniously rigged winches through the floorboards of their trucks that ran off the engine's transmission. By looping the rope around the winch's stem, they were able to slowly drag the great weight from the blue Atlantic. But not without constant adjustments, for the arc of the net and the fish inside were usually pulled sideways along the beach by a powerful current. And so the men needed to finesse the net down the shore through the breaking waves while slowly retrieving it, but also slowly reducing its gape.

If this tricky endeavor went well, stripers, bluefish, and almost anything else that swims in Long Island waters would begin to show up partway during the haul back, hanging from the mesh by their gills. But the great majority would come to the beach as a mass of life, splashing wildly with fins sticking out through the bulging netting of the gear's cod end. Sometimes the catch was so voluminous that the order was given to open the bag, letting hundreds of striped bass and other fishes swim free. Given the enormity of the net, few hauls were without any stripers at all, and we'd tag them, take tissue samples, and work them over for all kinds of diagnostic information, before walking each of the ocean-bright, mostly five-, ten-, and twenty-pounders to the surf to send them back on their Atlantic journeys.

One morning we brought in a haul of striped bass from the surf; as Vecchio handled them, he pointed to one and said, "That fish is going to beep." Moments later he passed a "wand" over it and it did beep. I was impressed. Vecchio wasn't psychic; he'd simply noticed that individuals with severely "broken" striping patterns almost always seemed to be hatchery-generated. This was eight years after striped bass had begun being stocked in the Hudson River and six years in Chesapeake Bay. Striped bass from both estuaries had been tagged with tiny coded-wire tags that were implanted in the cheeks of the little fish before they were released. These individuals were detected among a catch of stripers by waving a special electronic wand over the head of each fish and listening for that sound.

Completely intrigued, I proposed we launch a study of this striping phenomenon and other differences between the wild and the stocked stripers. In the end we scanned more than 3,300 striped bass over two seasons, and 102 tripped the electronic detector. There were other unexplained differences between wild "reference" fish we obtained from the Chesapeake and hatchery-bred Chesapeake fish, with the hatchery specimens having statistically significantly longer heads, shorter upper jaws, and longer trunks, among other contrasts.

What Vecchio had noticed casually about striping patterns held up. We used a simple scoring system for stripedness, dividing each flank into six sectors and then coding a zero for no broken stripes in a sector and a one for at least one broken stripe. Relatively few wild reference fish from the Hudson, four Chesapeake tributaries, or the Roanoke River in North Carolina scored a zero for the left flank, showing that perfect striping patterns occurred naturally but weren't common, with a small degree of broken striping with a score of one predominating. But for hatchery-reared stripers, the most frequent score was five, with zeroes absent, many twos through fours, and even a few sixes. The mystery remains; the parental broodstock for these fish were not selected for striping pattern, yet something in the hatchery environment brought out this tendency toward an almost checkerboard appearance, an outcome perhaps not important for the fish's survival but something that demonstrates how little we actually know about how rearing fish en masse in remarkably artificial conditions affects them.

Maybe more importantly, only 3 of the 102 captures of hatchery-reared striped bass were from the Hudson River releases, despite the numbers stocked into the Hudson being about half that of those that went into Chesapeake Bay, and the fact that the Hudson is considerably closer to eastern Long Island. It wasn't clear whether the Hudson stripers residualized in the river, or whether some abnormalities noticed among the fish propagated in the Hudson hatchery caused lower survival, but it seemed that an expensive program was providing little benefit. Nor was it clear that the Chesapeake hatcheries were needed. By 1995 the brief era of Mid-Atlantic striped bass hatcheries was over. It seemed in the end that little harm was done, but also, little good.

Hatcheries did play a significant and somewhat destructive role in the long-term survival of the unique striped bass stocks of Gulf of Mexico rivers that live tenuously as cool-water denizens of a warm (and warming) environment. Unlike the great migratory populations of the Mid-Atlantic, Gulf stripers rarely venture from rivers to the tepid coastal waters, and they sometimes are forced to gather near cool-water springs to survive the heat of summer. While damming and overfishing eradicated many populations in rivers westward along the Gulf, a relatively large stock persisted in Florida's Apalachicola River system. When the Jim Woodruff Lock and Dam was completed on it in 1957, so many stripers were caught by anglers in its tail-race that many were destroyed as a nuisance. Between this sad waste and the blocking of the spawning migration, a decline began that was answered by stocking, but naively, the source of the broodfish was not from Apalachicola but instead easily obtained striped bass from the Atlantic coast, where they were already being cultured—specifically, a generous 1.8 million fry and 125,000 juveniles and adults between 1965 and 1976.

Though Atlantic stripers were viewed as an expedient solution then, by the 1980s, through the lens of modern conservation biology, the stockings and consequent dilution of the native striped bass gene pool were seen as a major faux pas. It was more than likely that the Gulf race was better adapted to the high-temperature environments of the southern states. The challenge was to find some way to identify the remaining native stripers to selectively breed them. Luckily, some simple but fundamental research from the 1950s had shown that the average number of scales along the sensory lateral line that runs from behind the gills to the tail was higher for Gulf than for Atlantic Coast striped bass, and with little overlap in counts between them.

Sampling of striped bass in the Apalachicola from 1978 to 1982 showed barely more than a relict population remained of between 1,500 and 2,500 adults, made up of an estimated 43 percent native Gulf fish, 51 percent introduced Atlantic fish, and 6 percent a mix of the two. Now the question was, how to undo a genetic wrong? It seemed apparent: simply breed high lateral-line scale count stripers. But early on there was a surprise; their offspring displayed lower scale counts than the parents, showing less genetic and more environmental influence on what was thought to be a reliable diagnostic marker. To the rescue came the molecular revolution; each year techniques for characterizing DNA were becoming faster and more powerful. My colleague Isaac Wirgin, who had performed early DNA-based work on striped bass for his doctoral dissertation, was recruited to try and unwind the two strains at his lab bench.

Fortunately, the long isolation of Gulf stripers from their Atlantic counterparts had resulted in a mutation labeled *XbaI* that Wirgin found in 55 percent of the higher-scale-count striped bass and was absent among all fish screened from Atlantic rivers. But it also wasn't found anywhere else among the other Gulf Coast populations, as was true for subsequently discovered DNA markers, supporting the idea that the Apalachicola harbored the only native genes. Wirgin also applied some historical ecology, analyzing the DNA of museum specimens that predated Atlantic strain stocking, which showed a low level of introgression of genes from Atlantic fish. And so the managers had a tool, and the hatcheries geared up to breed only *XbaI* broodstock to try and recover the native Gulf of Mexico striped bass race.

Were they successful? Only partly, once again demonstrating how much better it is to prevent a potential problem than to remedy a mistake. Today striped bass management along the Gulf Coast is a hodgepodge of concessions to the realities of the striped bass stocking history and the degradations of Gulf rivers. The Gulf States Marine Fisheries Commission manages Gulf rivers for general striped bass fisheries, striped bass put-grow-and-take fisheries, Gulf race put-grow-and-take fisheries, and self-sustaining Gulf race populations.

Where has almost a century and a half of stocking left the anadromous fishes of the Atlantic? Severely depleted, despite these efforts. For striped bass, little effect along the Atlantic and a muddle along the Gulf. For salmon, virtually nothing to show for a half-century of efforts in the Connecticut River, and little progress elsewhere. For shad, not counting the many stocked by state governments, from 1872 to 1949, the federal government alone stocked more than *4 billion* American shad larvae. Most shad rivers did not receive ongoing shad stockings; nonetheless, those that did—either then or afterwards, or both, such as the Susquehanna—today have pitiful numbers of shad entering them each spring.

Hatcheries aren't the answer.

Chapter 17

Dam Removal:
Fish vs. Ignorance and Inertia

Every morning when I awake I ask whether I should
write or blow up a dam.
—*Derrick Jensen,* Disorderly Conduct *(Winter 2000–01)*

It was time to meet the guru of East Coast small dam removals, Laura
Wildman. Wildman, a tall blonde with a direct style, cheerfully greets me
with her left arm in a sling—injured while inspecting the face of a slippery dam. I've asked her to show me a few dam-removal projects in central
Connecticut, but my main agenda is to understand why she is so successful
at getting so many of these simple-in-concept but difficult-in-reality projects done.

Within minutes we establish that we both have an interest in old
houses, antiques, and "old wood," so we detour to her office to look at
her museum of finds made at dams. On the way Wildman explains that
because it was easier to build a new dam on top of an old barrier, modern earth and concrete dams occasionally contain remains of original dams
built with old-growth timber, coveted by musical instrument makers and
artisans, and sometimes salvageable. At her workplace, in a case on her
wall, are early wooden, iron, and brick artifacts she recovered at dams.
But my eye can't help but catch an outlier among them: a bright yellow
plastic carp. Wildman says that when the Anaconda Dam on Connecticut's
Naugatuck River breached partly in a flood, she was part of a team that
followed an emergency order to remove the rest of the dam. As they began
to restore the natural flow, almost cosmically, the yellow fish toy appeared
and passed downstream.

Wildman's office also contains framed photos of great western dams, which seem overly respectful for someone so ardently anti-dam (i.e., those with little societal value). But dams are in her blood, albeit with an attitudinal mutation: Her grandfather was a water resources engineer who helped build the Hoover and Roosevelt Dams, helping, she says, "to put Phoenix and Scottsdale in places where they never should have built a city."

Our first field site is Roaring Brook, a partly collapsed early stone dam that needs some TLC. Behind it, though, is a newly "released" stream that flows alongside a ten-foot-tall bed of stationary sediment that accumulated in the 200 or more years since the dam was built—a clear exhibition of Wildman's conviction that "a river wants to move sediment as much as water."

Laura Wildman showing artifacts of dam removals, including plastic carp that "passed" through the newly opened Anaconda Dam on Connecticut's Naugatuck River
John Waldman

Next stop is the site of the Zemko Dam removal on the Eightmile River. The river now flows freely through a gap, but Wildman looks at the remediation with the practiced eye of someone who has been involved in more than 125 dam removals. She's glad the project was completed, but she points upstream and says, "I don't like the ponding behind it; it's too extreme." This take speaks to the notion that dam removal is a mixture of art and science. Wildman's philosophy is one that essentially reduces to "The river knows best." She says we should aim to release a river "just enough to let it be what it wants to be, to take its handcuffs off." We travel to the site of another removal on the Eightmile. Wildman challenges me with identifying where the dam had been placed. I can't do it. For this river, at least, one reach has been healed.

I asked Laura Wildman how a leading dam-removal guru was birthed from a dam-building lineage. Not surprisingly, it involved an epiphany.

Wildman had a fascination for timber bridges and wanted to be a civil engineer. This sometimes-ebullient extrovert ground her way through an engineering degree, but found it dull and constraining. Her first job as an engineer, working on stormwater management in Seattle, was even worse; everything was done by the code, with no room for creativity. Dissatisfaction there sent her to the East Coast, still on stormwater but at a consulting firm, Milone & MacBroom. One new project involving diversion for drinking water put Wildman on Connecticut's Shepaug River to survey flows. Modeling hydrology and hydraulics on a pastoral New England river suited her. And now, Wildman needed to balance the needs of people and fish in ways for which there were no codes. In fact, there was a dearth of information and operating principles. Wildman would now need to think like a fish, and she embraced that challenge.

Wildman went to her boss and mentor, Jim MacBroom, and said, "This is what I want to do." He replied, "We all love these projects, but they don't come around too often." Fortunately, Connecticut, a leader among states in fish passage, had decided in the mid-1990s to reopen the entire main-stem Naugatuck River, a once highly industrialized and famously polluted system that wound through now mostly economically challenged cities that manufactured brass, such as Watertown, and Naugatuck, the latter with a dubious claim to fame: This is where Naugahyde was first produced. Wildman sensed her opening and put yeoman effort into responding to the state's Request for Proposals, including her suggesting the construction of "nature-like fishways," then a still-novel concept in the United States. Her company won the contract, dams began to fall, and Wildman established her reputation, both at her firm and, later, as chief engineer at American Rivers, a progressive national nonprofit institution dedicated to celebrating and restoring rivers.

The Naugatuck restoration will be a mix of removals and fish-passage projects on its eight dams. Five dams have been removed. A ladder on the Kinneytown Dam, lowermost in the system, had about 25 fish species using it. In 2012, 57 American shad, 28 alewife, 35 sea lamprey, and 20 sea-run brown trout were counted moving past the Kinneytown Dam, and unprecedented numbers of newly spawned fish were seen passing back down. Not exactly a plethora, but these are not decreases from a happier status quo, but rather the encouraging first pioneers in a system just taken back from industry after two centuries of abuse.

Former Interior Secretary Bruce Babbitt during the Clinton administration was a fanatic for dam removal, and an astute observer of the trends and ironies involved in the process. Babbitt not only presided administratively over numerous removal projects, but ceremonially, too, graduating, as he put it, from sledgehammer to jackhammer to wrecking ball, moving to sky crane, and even to C-4 explosives. Babbitt noted in 2002 in an article titled "What Goes Up, May Come Down," that once upon a time, dam opponents were the true fiscal conservatives, urging caution, pointing out that dams typically cost taxpayers more than estimated and that alternatives to dams exist, and calling for environmental impact studies. Meanwhile, dam builders pressed for rapid change under a "trust us" philosophy: "[W]ith all the benefits their dams would surely provide, all other concerns would be sorted out . . . once the dam was under way." Dams were the panacea for floods or fire, voltage or storage. But over the years, concrete crumbled, reservoirs evaporated or filled with sediments, fish runs were blocked, and fixes were costly. Slowly, dam removal became the answer—a major means to repair degraded rivers.

Now, according to Babbitt, it's the dam owners who are preaching fiscal austerity, blasting some removals as too expensive, claiming that the local title-holder and recipients of the benefits of dams should not have to finance their deconstruction, demanding compensation for displaced energy operators or other water users, and arguing for other options to be pursued. And, ironically, they are calling for expensive and time-consuming environmental impact studies about the potential impacts of dam removal in the face of any uncertainties. Now the dam-removal lobby is saying, "Trust us—taking down dams is the answer."

In Babbitt's view, change has arrived; the heyday of dams has come and gone. In the 1990s the public asked, Why? Or whether? In the new millennium they ask which ones, when, and how? Yet Babbitt as Interior Secretary was not a wide-eyed zealot. Though removing dams in his opinion often makes ecological *and* economic sense, the impact of deconstructing a dam needs to be carefully evaluated before removal and objectively considered afterward, even, and especially if, examination of pre-dam and post-dam construction was never conducted when the dam was built. Babbitt also believed that dam proponents should be recognized even by—especially by—those same dam opponents who were excluded from past decisions to build.

Fairness taken to an extreme? Why did Babbitt hold dam removal to a higher standard than dam construction ever faced? Because he believed that if such concerns go unanswered, then the future of dam removal may eventually erode to become as vulnerable, unstable, and obsolete as some of the dams it will erase. Babbitt did not want to replace the old way—*build now, ask questions later*—with *remove now, analyze outcomes later,* but instead, to recognize the socioeconomic and ecological complexity of the issue.

Babbitt's philosophy was forged in the extreme heat of congressional oversight, influential power utility companies, agricultural interests, and various federal agencies. Wildman works at the lower-stakes but still often-contentious level of state agencies, private dam owners, and communities.

These differences don't mask the obvious commonalities—that at both extremes, a large mass of material placed by humans at an earlier age in many instances is continuing to perform ecological harm to otherwise-productive rivers through a combination of ignorance and inertia.

Wildman shows me a final site on the Eightmile, Ed Bill's Dam, a major rock-and-mortar structure that holds back a respectable-size pond. The barrier already has an Alaska steep-pass fish ladder that doesn't perform well. But the dam-removal issues here are mostly in the social realm, not in the technical.

After viewing it we begin to drive away, but Wildman launches into her philosophy on getting the job done. This wisdom distilled from a career performing this arcane endeavor is too valuable to commit to memory, so I pull my car over to take notes. Wildman, now director of the New England Regional Office of Princeton Hydro, says to begin with a dam's history; it's the context you'll be working in. Survey the infrastructure of the dam and around the impoundment. Look at the sediment held behind the dam—will it remain in place or be mobilized? Which fish species occur now, and which are targeted for restoration? Are there other goals for the project, such as education, recreation, or water-quality improvement? What is the dam's configuration, and how will heavy equipment access it? Will there be a need for active channel restoration once the impoundment is drained? And will the community need to be engaged or convinced of the project's worthiness?

Wildman speaks with the confidence of experience when she says that after the first field day at a new site, she and her team know exactly what to do. For the next year or two, the rest is red tape and social issues. At the end of that time, as the actual removal is about to begin, by her estimate, 90 to 95 percent of what will occur is what they visualized on that first day.

It seems as if most streams in Massachusetts have one of two prefixes in their names: "Herring" or "Mill," representing the two battling historical human uses of those flowages. I'm on a Mill River in Taunton this autumn day, visiting the removal of what is known locally as the State Hospital Dam, a legacy barrier too minor to have been bequeathed a grander appellation. Minor in societal terms, but absolutely major to the ecological health of the Mill River. The river once was a prodigious producer of river herring, but these were choked off as its waters came to run through some seventy mills. A historical account of the Taunton River, fed by the Mill River, soon after dams cut off the alewives is plaintive: "The cry of the poor every year for want of the fish in the Taunton is enough to move the bowels of compassion in any man, that hath not a heart of stone." Now, a partnership of federal, state, and local agencies, businesses, and NGOs have come together to unblock the Mill River and return herring to their historic home. Ultimately, three dams will be removed, and a fish ladder will be built at a fourth. When the restoration effort is complete, migratory fish will have access to more than thirty miles of high-quality tributary and main-stem habitat. The State Hospital Dam was the first of the three dams to be removed.

My tour guide is Beth Lambert, a River Restoration Program manager with the State of Massachusetts. She surprises me by bringing along Mike Bednarski, my fish-crazy former master's student, recently hired by the Bay State to help restore diadromous fishes after completing a doctorate working on sturgeon in Georgia. Lambert also has serious river cred, a degree in geomorphology, and fifteen years working on rivers in Oregon, Alaska, New Hampshire, and Massachusetts. The project we're looking at began more than two months earlier, but just two weeks earlier, it was celebrated publicly with a ceremony. I wished I had been there; the most prominent attendee was the refreshingly outspoken local congressman, Barney Frank, who received praise for securing federal funding for the bulk of the $800,000 undertaking. Frank is famous for his wit and his impatience, memorably telling one woman at a town hall meeting who likened President Obama's health-care plan to Hitler and Nazi policies: "Ma'am, trying to have a conversation with you would be like trying to argue with a dining room table. I have no interest in doing it." Apparently he was more mellow on this day, calling the project a refutation to those who typically disparage public-private partnerships, and getting a laugh when he said considering the number of parties involved he "would have bet heavily that this never would have happened."

It's a Friday afternoon, and various forms of heavy equipment are being put through their final manipulations of the terrain before shutting down for the weekend. What lies before us is the bed of the five-acre lake that sat behind the dam. But now a river runs through it, and in a strange manner. The main channel is sinuous and natural-looking, but it was not created by nature; its curves have been intelligently imposed to befit this particular river gradient. This course is encouraged by webbing that secures the banks in place, ready to receive native plantings that will provide habitat and hold the banks together. But along one bank is another, smaller stream, outlined by its flowing through a plastic sheet–lined channel. This is the diversion conduit, built so that the main flow could be routed through it so the spillway could be lowered. One patch of the former lake looks scraped flat; Lambert explains that contaminated sediments were carted off to an upland site nearby. Before we head downstream to where the dam was located, I spot a mature couple sitting in a homemade plastic sheet and lumber canopy in their backyard, watching the work proceed. Lambert says, "Those are the Machados; they love this project." I decide I need to meet the Machados.

Not far from the dam on the opposite bank is an auto repair shop whose owner would have preferred to keep the pond behind his parking lot. But the dam, gone only weeks, is already a distant memory; looking at the river, I can't tell where the dam sat, but Lambert's photos show the progression of its removal and how it once fit into the landscape in front of me. Bednarski, who's unrepressed enthusiasm (for anything fish-related) would make him a golden retriever if he was a dog, tells me excitedly that he can't wait to monitor the newly opened system; he thinks it might support 200,000 river herring. Bednarski is especially excited about the arc of history; while researching the river he'll be working on, he found a quote from 1920 by a fishery biologist who concluded that it was so badly polluted by industrial

Joseph and Robin Machado in their backyard dam-
removal and stream-restoration viewing canopy
John Waldman

waste and obstructed by dams that "the reestablishment of the old herring fishery is an impossibility." Bednarski, with the still-uncorroded ardor of a newly minted PhD, smiles at the thought of doing the impossible.

We walk around the lake bed and surprise Joseph and Robin Machado, who are happy to tell me what the project means to them. A lot, it seems. They are out there watching the show every day, no matter what the weather. They are familiar with each detail of the work, they know the crew members by name, and they've saved the eighty-two turtles they've found by relocating them to other sections of the river. Joseph worked at a factory up the street and has lived here long enough to remember when the river was nearly lifeless. In recent springs he's watched the frustrated river herring butting up against the dam. To him, "It's a dream come true."

As I drive away, I think about how Wildman considers the ecological rationale for taking down each and every dam that serves no clearly useful purpose anymore—that such "defragmenting" allows a river to be dynamic and productive. And how she went further, arguing that "People fought these dams when they were being put in for the greater good. Now we have our entire industrial past rotting in rivers. If the industry is gone, why keep the dam?"

I also am struck by two conflicting thoughts. The first is just how prosaic a dam removal is: the engineering plans, the temporary rerouting of flows, the breaking apart and removal of concrete, the moving of earth. The second is just how marvelous it is: the healing of an old industrial scar, the reconnecting of long-isolated biological communities within different reaches of the river, the rekindled linkages between fresh waters and the sea. For the Machados, it is a dream come true. For the ecology of a river, nothing less than the righting of a wrong.

Chapter 18

Fish Passage, or Not

Even the upper end of the river believes in the ocean.
—*William Stafford*

Turning off onto Route 154, a two-lane highway along the Connecticut River in central Connecticut, I immediately see a sign for "boned shad," and am pleased to know the ever-declining fish is still relevant to the towns on the river's lower, undammed reaches. I then overshoot the Shad Museum, which is easy to do. The museum exists in a backyard shack that happened to come along with some commercial property bought by Dr. Joseph Zaientz—a bona fide shad fanatic. Zaientz, who suffered a motorcycle accident decades earlier, greets me cheerfully, hobbles through the parking lot, and settles into a chair, cuing up an old shad-fishing video he thinks I should see. But in walks a netter, John Smoloski, who put three decades into shad fishing, supporting himself the rest of the year as a carpenter. A few minutes later, an angler shows up carrying two coffees, one for Zaientz. Dave Roberts, a young man with his own version of the shad bug, generously helps care for Zaientz and his grounds. I can't believe my good fortune; I've wandered into a perfect confluence of shad aficionados from different and maybe even competing interests—historical, commercial, and recreational—and I realize the shack is as much a shad clubhouse as a shad museum.

Everyone is ready to talk shad. We consider the nature of the American shad, the state of the river, and how the species is faring in the river. The importance of fish passage. How, unlike for the murky Hudson, netters have to fish at night so the fish won't avoid the mesh in the clearer water. The consensus is that it is an up year for shad in the Connecticut, though nothing like the good old days. But the roes, or female shad, were on the small side. Stripers were down, which might be good for the salmon; Smoloski had opened striper bellies in the past and found salmon smolts in them. But no one had seen any adult salmon this year. The sense was, they never would; the restoration was doomed. Smoloski said it would be nice to catch a salmon, but frankly, "It ain't going to happen."

Zaientz remains set on playing the video, but Roberts provides a rundown of the pros and cons of his semisecret local shad-angling spots. Meanwhile, I scan the room—a solid collection of random but relevant artifacts: lures, commercial gear, books, signs, and shad derby buttons. And the pièce de résistance—a huge chart made by a netter of the commercial fishing heyday decades before, that shows the various stretches of river where the shadders once drifted their nets: the Maramos, Higganum, Rock Landing, and Salmon River Reaches. The chart also reveals the precise locations where netting occurred, the names of the fishermen themselves (the most noteworthy: "Curly Nosal"), their favorite spots, old photos—a trove of historical information fortuitously and endearingly captured in homemade fashion before it was lost, a treasure in time recorded on a simple slab of cardboard. Looking further, at a large metal net-buoy with a torch fitting hanging from the wall, I can't help imagining a quiet May night with small fleets of skiffs drifting these reaches, the ends of their nets marked by flames in the darkness as the shad drive headfirst into the twine.

Finally, Zaientz loses all patience and presses the ON button, playing an old-fashioned straight-up report on the river's shad fishery from back in the day. Smoloski and Roberts are good sports and watch it for the umpteenth time, with Smoloski again seeing himself as the star, a younger man, but then, as now, drifting the nets all night, then staying awake long enough to bone a few dozen shad and get them to a distributor before catching a nap and repeating the cycle again—hard honest work that services the soul more than the wallet. I ask Zaientz if he's a boner, a term that often elicits chuckles among the non-cognoscenti. His answer astonishes me. Zaientz, a former dentist, a man at ease in a patient's mouth, who could tease a dead nerve from a snaky molar root, said, "I could never do it." It's that challenging.

What is the future of commercial shad fishing on the Connecticut River? And of the museum itself? Smoloski says there are about eight fishermen left, where once there had been eighty. Why? "It's cold, windy, and rainy out there. The younger generation isn't picking it up." Will the museum survive in the long term once Zaientz can't volunteer his weekends anymore? And what kind of museum will it be if it does—a small labor of passion to a tradition finis, or a synopsis of what has taken place from the beginning to the present day, of a long history that will carry on? What happens upriver with the river's dams will be instrumental in answering these questions.

Later that day I travel upriver to Holyoke, Massachusetts, a mill town that hasn't aged well but which once flourished making paper, powered by a dam and canal system that harnessed the energy of Hadley Falls, a sheer but once passable rapids—at fifty-three feet, the Connecticut River's largest natural drop. My destination: the fish elevator at Holyoke Dam, owned by Holyoke Gas & Electric (HG&E). I soon learn that HG&E not only owns the dam—perhaps not on paper, but for all intents and purposes—it also owns the river's shad. The Holyoke Dam holds back a reservoir, and with the heavy spring rains of 2011, a thousand-foot-wide sheet of whitewater pours over the dam and humps over the rapids below.

In the parking lot HG&E has set up an open wooden hut where the weigh-in for the company's annual shad derby occurs. There is still more than an hour to go, and its attendants only intermittently need to weigh someone's catch. At the door of the hydro facility, an HG&E staffer is there to deflate visitors' expectations. Because of high water, no fish are being lifted today; in fact, no fish have passed this barrier for the past six days. Nonetheless, a stream of visitors comes to read the signage and to watch a video about the dam, and to see the turbine room, the fish-lift mechanism, and the now-empty fish-viewing window. I pause outside and feel the generic sensation of a large hydropower facility, something perceived both with ears and body, that of barely restrained power. The sheer force of the flow is manifested in a symphony of roars and hisses from the river itself, which doesn't quite harmonize with the steady thrum of the spinning turbines. I have the uncomfortable feeling that if a rivet popped, the whole enterprise would blow.

Inside is fairly standard utility company propaganda. On one wall is the generic salmon-boosting poster representing the great coalition of federal and state agencies. Its motto: "A Healthy River Makes Healthy Fish." No argument there, but I don't see the connection to the Connecticut. Below that is a vintage image of a leaping salmon in a pristine setting; I consider whether a salmon taking a ride in a metal tub up the face of a dam would be more apropos.

Signs line a cul-de-sac on the first floor explaining the importance of the facility and the success of the fish elevator. Forty-five megawatts of power emanate from the two main turbines and some smaller units in the canal. Fish have been lifted since 1980. In 1980, 376,757 shad took the elevator past the dam. Shad peaked in 1992 at 721,755. In 2010, 164,439 made it over. Salmon counts were 119 in 1980; they peaked in 1992 at 368 and were 41 in 2010. Blueback herring were faring even more poorly than salmon; they were most numerous in 1985, at 632,255; from 2006 to 2010, they numbered 21, 75, 84, 40, and 76, respectively.

A TV monitor ran a loop from a short news piece on the fish lift, and also called the dam and fish lift the "birthplace of innovation." The problem was, the innovation I saw was largely jury-rigged. To spill enough water at the right location to attract migrating fish under low-flow conditions, much of the top of the dam had been retrofitted with a rubber bladder. HG&E's signs tout various improvements they made to the fish-passage facility in 2005. Some seem to matter more to the engineers than the fish, like the installation of a centralized electrical system. Some matter more for public relations, such as a widened fish-viewing window for visitors. Others seem like potentially useful tweaks: a single queuing system for fish instead of two and a fourfold increase in fish capacity. And yet, scrutiny of the before- and after-2005 results revealed little difference in numbers of fish passed.

I climb the stairwell, glance at the empty fish-viewing window, and then step out on a tall platform over the river. Through my shoes I feel trembling as the entire apparatus strains to maintain its footing against the power of the flow; the nearly unfathomable force is distinctly palpable even from high above. Later, when I am almost out the door, I notice that one of the signs puts the electrical power generated into perspective—in layman's terms. HG&E proudly announces that those 45 megawatts equal the draw from 300,000 100-watt lightbulbs. Illuminating. (Pardon the pun.) But this factoid leaves me less than impressed; indeed, I'm almost embarrassed for the utility. The ecological health of a river is being compromised for what appears to be a truly modest amount of electricity.

By 4:30 the crowd has grown at the HG&E shad derby booth, everyone awaiting the 5 p.m. weigh-in deadline and the announcement of the winners and their awards. Anglers whose catches did not make the top ten in the senior or junior categories did receive a 46th Annual Shad Derby T-shirt as a consolation prize. Although I am clearly shadless, an HG&E employee manning the booth sees that I am serious about viewing the event and tosses me a T-shirt. On a personal level it seems that the HG&E visitor center tour guides and the men in the weigh-in booth are dedicated employees and decent folks. But I nonetheless have an overwhelming sense that something is wrong here: Between HG&E controlling the very migration of shad past their great barrier, and their promotion, sponsorship, and conduct of the shad derby, this utility, which exists only to generate electricity, appears to have sovereignty in effect over this and other important fish populations in the river.

Over the final few minutes, boaters and shore anglers rush their catches to the scale, most tipping it at four pounds and a few odd ounces. Shortly after 5 p.m., a Springfield fireman, David Bello, is awarded the $1,000 senior division grand prize, with a shad weighing an even five pounds. The fish had been caught and weighed in the day before, David saying, "I sweated through this; I thought, 'Five-zero ain't going to make it.'" It was the smallest among the winners listed on the derby board through the years, far less than the eight-pound, eleven-ounce winner landed in 1985.

The shad derby does help to maintain a local connection to these fish, but not in the traditional sense, as an esteemed food source, but instead, through a brief annual competition for monetary prizes. As the other award winners are given their trophies, I notice a box of shad off to the side of the booth. Many *Alosa sapidissima,* "the most delicious of herrings," were left behind by fishermen after being weighed, their marvelous table possibilities ignored. They lay there sadly, stacked in a heap, mouths agape, the violet and gold iridescence of their scales fading in the afternoon sun.

The next morning I station myself in a park near a canal at Turners Falls, another seen-better-days Massachusetts town on the Connecticut River, not far from the Vermont border, and await a local river conservationist, Karl Meyer—although *river gadfly* would be another description. Meyer is the one-man conscience for the good of shad in this reach of the Connecticut—not as a voluble voice at public meetings or a ranter in the press, but as a persistent, thoughtful advocate for rethinking the current (and long ongoing) management scheme for the river's diadromous fish. I like him immediately, finding he evidences a passion and an independence that seem slightly Thoreauvian. Indeed, he leads a simple life, earning just enough by writing and driving a school bus and performing other odd jobs to allow free time to wage this battle.

Meyer's tour begins with the canal, a trough used by most shad that attempt to surmount the Turners Falls Dam to the final, currently somewhat accessible reach of river. Meyer points out the obvious: This is a highly artificial and challenging route through which to expect fish to find their way. He and others believe it is an "ecosystem death trap," essentially a two-mile break in the biological continuity of the river. Indeed, in most years far less than 1 percent of the shad that enter the lower Connecticut River make it past this dam. And if they do, they may become confused inasmuch as the gigantic Northfield Mountain Power Station, a pumped storage facility built into the hillside above the reservoir above the Turners Falls Dam, sometimes draws so much water that it causes the flow to reverse and head *upriver.*

Next we look at the stretch of natural river immediately below the dam, a wide rocky bowl, one so fetching and obviously strategic for fishing and associated purposes that you could just about feel its history. Indeed, its known past is lengthy and, unfortunately, infamous. It was here in this pool, below a passable forty- to fifty-foot-high cataract, known as Peskeomskut (Great) Falls, that Native Americans of the Pocumtuk Confederacy captured shad and salmon for ten millennia, using the banks to camp and cure the fish. It is also where, in May 1676, 150 colonists led by Captain William Turner launched a stealth attack early one morning on the Indian encampment to avenge the theft of cattle some days earlier. The savage onslaught killed several hundred men, women, children, and elders; many were shot, and some drowned in the river, trying to escape. But neighboring tribes became aware of the skirmish and closed in. On the confused retreat, nearly forty of the white men were killed, including Captain Turner, who lent his name to the site.

Meyer later takes me along a path through the woods about a mile downriver. There, at a deep hole, one of several anglers is busy fighting a shad. This delights Meyer; the thrust of his one-man crusade is that the Connecticut is one of the great shad rivers, its enormous potential being squandered on an effort to bring back salmon, something he views as a quixotic and misguided quest. He sees now as the time to end the salmon program. Hurricane Irene has largely destroyed the major Connecticut River salmon hatchery at White River Junction, Vermont, with $10 to $14 million needed to rebuild it; it also potentially contaminated the fish there with a slimy invasive algae often referred to as "rock snot." To put it in especially stark terms, he points out that in 2011, salmon represented less than three-hundredths of 1 percent of all anadromous fish returns, while devouring 90 percent of all funds for migrants. By Meyer's reckoning, each returning adult salmon in 2011 cost taxpayers between $110,000 and $264,000. This is roughly what they would cost if they were made of solid gold.

Driving down the Connecticut River Valley of Vermont on a bleak, early-winter day, I stop to look at the canalized and dammed river at Bellows Falls. The fish-passage facility is, of course, closed this time of year because the anadromous fish aren't running, but I think of all the resident fish that might want to move up- or downriver and can't. And I also wonder: Just when did an anadromous fish's fundamental migration up a river became "passage"?

Dams, and all the problems they pose for movements of riverine creatures, have been part of human history for four millennia, at least since the
ancient Egyptians built the Sadd el-Kafara somewhere between 2950 to
2750 BC. This was an ambitious but flawed structure. Almost 45 feet high
and nearly 350 feet wide at its crest, the dam had rubble masonry walls and
was filled with 100,000 tons of gravel and stone. It's safe to say it was not fitted with fish-passage facilities. It also did not block any fish movements for
long; it failed after a few years. The ancient Egyptians did a far better job
with their pyramids. Ironically, four millennia after Sadd el-Kafara, a pro-
dam book was written titled *Dams: The Useful Pyramids.*

Though dams no doubt were created in deep time prior to Sadd el-
Kafara, and hundreds of thousands were built worldwide following it,
attempts to get fish past these barriers are only recently documented. Rough
fishways made from bundles of branches are known from the seventeenth
century in France. A fishway was patented in New Brunswick, Canada, in
1837 by a mill owner. An early sophisticated fish ladder was documented
on the Pawtuxet Falls Dam in Rhode Island in 1880, and others elsewhere
in the United States perhaps a decade or two before that. Today fishways
are used worldwide, but at low frequencies of application and with varying
degrees of success.

There is something of a hierarchy in implementing fish passage at a
dam. Most fundamental and most effective is the obvious: Take it down.
But if this is not going to occur for whatever reason, a desirable option, but
only on some low-head dams, is to cut a notch in the barrier to allow fish to
move through it. Then there are fish ladders. Fish ladders are metal, wood,
concrete, or combination chutes that have horizontal or vertical baffles that
create quieter chambers in which a fish may rest before blasting through a
pinched flow to the next chamber. Fish ladders work best when they are
short, which means they are rarely applied to huge dams. But they require
considerable space because the slope needs to be 15 percent or less, with less
being better. One solution is to construct the ladder with one or more direc-
tional switches, like in a staircase, but that complicates the passage.

In his typically elegant way, John Hay in his paean to the alewife spoke
to the compromise inherent in fish ladders. "There is no such thing, I have

been told by men who were in the business of making them, as a good or even adequate fishway. There is always an imbalance between the purposes they serve and the results." And: "They are built to try and bring back what man has taken away; though it should be said that they are as much in man's interest as the alewife's. Commerce is the main benefactor of their success."

Dams too large for fish ladders may be fitted with fish lifts, also known as fish elevators, such as the units at Conowingo and Holyoke Dams. Fish lifts have "attraction" flows that serve to gather anadromous fish in a trough that is then lifted to the top of the dam, where it opens to a tall channel that connects to the reservoir behind the dam. The channel usually has a viewing window for both technicians who count the fish and for the public. Fish lifts can work well at passing fish upstream; they don't pass any fish downstream again; and they don't work when flows are too high or too low.

The very names of these fish-passage stratagems ring oxymoronic when linked with fish; *ladder, lift,* and *elevator* are not terms of kinship for creatures with fins. Then, when all else fails, as a form of triage, is *truck and transfer.* Anadromous fish are collected below a dam and driven by a tanker truck to be released upstream of a dam. It saddens me profoundly that these beings—which evolved such majestic life cycles over millennia, were born in rivers and spent years in the utter wild at sea, and traveled perhaps hundreds of miles in queuing to their birthplaces—are finally forced to complete their journeys dependent upon the internal combustion engine.

Later that day I hear a pitch on the radio to keep open the Vermont Yankee nuclear power plant on the Connecticut River because it provides "clean energy." Even though the company's website features the mantra *Safe, Clean, Reliable,* in 2010 the Vermont Senate voted 26 to 4 not to relicense it. *Reliable* seems a reasonable brag; *Safe* can be questioned because of the general proposition that the splitting of protons is inherently risky, and because of the particulars: that in 2007 one of its cooling towers collapsed, and that in 2010 radioactive tritium leaked from the facility was discovered trickling from its underground pipes. But *Clean* is in the eyes of the beholder. The company touts the reductions in its acid-rain and climate-warming atmospheric emissions in comparison with similar-caliber fossil fuel plants. But no mention is made of its evident and undeniable harm to the river that provides the water that cools its generators, a level of ecological injury that obviates the notion of "clean."

The Vermont Yankee plant sits just behind the century-old Vernon Dam that holds the reservoir waters that cool the nuclear-driven turbines. The dam itself has a small hydroelectric operation now operated by TransCanada. TransCanada has invested considerable funds into upgrading the hydro plant, but Internet discussion boards about the site were rife with statements by anonymous employees about decrepit and unsafe conditions there. And the hydro plant has been certified by the Low Impact Hydropower Institute. The Low Impact Hydropower Institute has a yes/no-style template that must be addressed in the application. A consultant to the utility handled this. Although there was a level of frankness in the submission about fish passage, including the absences of downstream and eel-passage facilities, the fact that the dam's fish ladders were designed to pass 40,000 adult salmon and 750,000 adult shad annually helped it pass muster. The fact that only 19 shad and zero salmon actually passed the Vernon Dam in 2009, and that these numbers weren't anomalous, apparently didn't figure into this decision. And so a dam that produces only 32 megawatts of electrical power and holds back water for a nuclear plant about to be decommissioned and has fish-passage facilities for multitudes of fish but which receives almost none because of the poor fish passage downriver is approved until 2014, under the fantasy that it is causing only "low impact."

Since the 1960s operators of large hydropower dams along the Eastern Seaboard have been charged by the Federal Energy Regulatory Commission with restoring the anadromous fish populations they had helped to destroy, but by the first decade of the next century, little obvious progress has been made, and some restorations have even gone backwards. Yes, these dams were fitted with fish-passage devices, what some might term "halfway technologies": fish lifts on some, various forms of fish ladders on others. And yes, trickles of fish still came through. Each dam had been assigned some target for numbers of particular species, targets that likely were well below historical levels but which, if met, would have provided a modicum of angling and commercial fishing.

Colleagues and I felt it was time to investigate the progress of these restorations holistically. We selected three rivers as examples: the Connecticut, the Merrimack, and the Susquehanna. In the end our dismal quantification managed to surpass the feeling of pessimism shared by many. Historically, most of these rivers had Atlantic salmon runs on the order of tens of thousands, American shad runs on the order of millions of fish, and river herring runs on the order of millions, or more likely tens of millions. Goals for restorations on large rivers with hydropower dams are more modest, but are often in the tens to hundreds of thousands. Today, however, despite the implementation of these fish-passage technologies, many of these runs are reduced by 95 to 99 percent, with some suffering extirpation. An example is river herring, where mean returns to the Susquehanna, Connecticut, and Merrimack Rivers from 2005 to 2010 fell to relict levels of 79, 138, and 768 fish, respectively.

Fish-passage efficiency data for large Atlantic coast rivers showed that, despite state-of-the-art technology, these fishways have been largely ineffective at passing fish upstream to their historical spawning habitat. Of the fish that passed the first dam, on average only 4, 16, and 33 percent passed the second dam in the Connecticut, Merrimack, and Susquehanna Rivers, respectively. System-wide efficiencies, i.e., passage of the ultimate dam, averaged less than 3 percent. Lewis Thomas in *Lives of the Cell* defined halfway technologies in medicine as "the kinds of things that must be done after the fact, in efforts to compensate for the incapacitating effects of certain diseases whose course one is unable to do much about." We concluded that these fish-passage programs are prime aquatic examples of halfway technologies.

These halfway technologies make for truly onerous trips for fish that forge far upriver. Take a shad attempting to reach the limit of its ancestral migrations on the Connecticut River: After entering the mouth of the river from Long Island Sound, at more than fifty miles upriver it has to pass over the remains of the early wooden crib dam in northern Connecticut, at Enfield Rapids. Moving into Massachusetts, in the big pool at river mile 87 it must sense the attraction flow of the Holyoke fish elevator and then pass to the reservoir above. Next comes the fish ladder at the Turners Falls Dam at river mile 122. If it negotiates that dam, it swims into Vermont and is faced with a fishway at the Vernon Dam at river mile 142, and another at the Bellows Falls Dam at river mile 174. (Bellows Falls is the natural barrier that stopped anadromous fish from migrating any farther up the Connecticut River. One old-timer in 1865 recalled landing 20 salmon and 1,300 shad there with one pass of a scoop net. In 2011 only 46 shad could have reached Bellows Falls, because that's how many passed Vernon Dam below it.) If the shad spawns successfully, it has essentially no chance of reversing its way through these barriers and returning to the sea to become a repeat spawner. And any young it produces will need to somehow find their way downriver past various bypass channels, log sluiceways, or over the lips of the dams.

To arrive at the Silvio O. Conte Anadromous Fish Research Laboratory, you pass through the back streets of Turners Falls and proceed down "Migratory Way," a nomenclatorial assertion not necessarily supported by reality. Migratory Way follows the downstream section of the two-mile diversion canal below the Turners Falls Dam until it dead-ends past the lab, its water flowing through various fish-confusing slots and spillways that generate power and attempt to pass fish both upstream and downstream. The canal is a distinct mixed blessing for migratory fish. Most that reach this stretch of river try to ascend the Turners Falls Dam via the difficult canal route, which till now has been distinctly fish-unfriendly. But by providing readily available waters at a height to be dispensed in controlled fashion through the flumes and apparatus of the lab, the canal enables state-of-the-art research on fish ladders to proceed.

The installation owes its existence to Silvio O. Conte, a congressman from western Massachusetts and an avid outdoorsman. In 1965 he introduced legislation, the Anadromous Fish Conservation Act, to try to realize the great dream—to restore salmon to the Connecticut River. By 1967 the US Bureau of Sport Fisheries and Wildlife, the Bureau of Commercial Fisheries, and fish commissioners of Connecticut, Massachusetts, New Hampshire, and Vermont had assumed responsibility for the restoration and preservation of migratory fish. That included the inauguration of the Connecticut River Atlantic Salmon Commission. The laboratory was going to assist this effort.

My tour guide on this October day is Ted Castro-Santos, a friendly and fit-looking, late fortyish, highly enthusiastic researcher at the lab. Castro-Santos tells me he landed here professionally because of the pull of this land: Born nearby, he remains deeply attached to the central Connecticut River Valley. Though trained as a weasel-tracking mammalogist, he later applied for a "fish tracker" position with a consulting firm and became intrigued with following scaled creatures, earning his PhD while working at the lab.

We begin in several warehouse rooms. One chilly space holds circular tanks with research-ready alewives, salmon, trout, and sturgeon. Another has scale models of new-design fish ladders ready for wetting. No matter how right a highly engineered ladder looks on a computer screen, it doesn't begin to be honestly vetted till water flows through it—and even then it's nothing but a streamside eyesore if fish don't use it.

Next we visit a seventy-foot-long, glass-sided flume on the grounds outside that were used by Castro-Santos, his supervisor Alex Haro, and the rest of the fish-passage team to test the swimming abilities of various fishes. In fact, this was the subject of Castro-Santos's dissertation, and he becomes especially animated describing the experiments. "Burst speeds" were remarkable, he says; some species reached an astounding twenty-eight body lengths per second. But he also tested sustained swimming and the consequent thresholds of fatigue. Soon he is drawing graphs with his finger in the dew on the side of the glass. He loses me when he gets to calculus and derivatives, but the message I receive is that fish species behave differently, yet there are broad principles that apply to all. Castro-Santos brings his knowledge to a simple distillation when he says, "Fish ladder design is not physiology alone, and it's not fish behavior alone; we need both."

The showpiece is saved for last—the great flume that the Conte lab is known for. Although not inclined to the uber technical, I am wowed at its potential. The main concrete channel is 110 feet long, 20 feet wide, and 20 feet deep, and it is bracketed on each side by accessory flumes that are 10 feet wide. Castro-Santos had made sure that the current was turned on for me. Suffused with that eerie lighting peculiar to warehouses and humming with the heavy flow, it is a strange hybrid of infrastructure and nature, as if a watercourse had burst through a factory. We climb to the walk high on a terrace above the channel and peer down at a little tongue of the Connecticut River pushing through an experimental passage window that fish had been "challenged" with during spring. Castro-Santos says it had worked; even notoriously fussy sturgeon swam through it. Still, he acknowledges the limits of the system, too. "We are not a river here; it's best to ask simple questions." As I lean over the railing, I notice wide vortexes spinning on the

A fish ladder leading to the diversion canal
downstream of Turners Falls Dam, Turners
Falls, Massachusetts
John Waldman

opening's downstream sides, every boil and seam beautifully reflected by the lights, and I imagine that da Vinci would have loved this view, and would have asked his own simple but probing questions.

After lunch we visit the real world outside, the Turners Falls Dam and the diversion canal. Karl Meyer had shown me this spot during the shad run and had bemoaned the poor passage results. With Meyer, and again with Castro-Santos, it was obvious that the canal in the area of the dam is hell for a fish. The flow shoots through the channel at a frightening speed. Whitewater kayakers would be happy here. But it's not even moving full bore. The dam's superintendent comes out to where we are standing over the river and tells us that it's jetting at about 8,000 to 10,000 cubic feet per second, just two-thirds of its maximum.

As I take in the dimensions of the dam, the canal, and the fish ladders and their connections, I realize the shad are being asked to make migratory sense along "Migratory Way" of a two-mile-long, jury-rigged obstacle course. Normally, a fish ladder or lift would be built directly on the face of the main dam, but the first version of the diversion canal was constructed for navigation in 1798, was reconfigured for power production in 1869 (Turners Falls was later known as the "Home of the White Coal"), and the present fish ladders are retrofits. To reach the fish ladder tucked into the east corner of the canal that will actually get them over the Turners Falls Dam, about 10 percent of the shad that make it this far ascend from the natural river bed near the base of the dam, up a ladder sixty-seven vertical feet, into the canal. Most, however, attempt to enter the lower end of the canal at the "Cabot Fishway." To climb there, shad enter the ladder along a short, straight run, make a right-angle turn in a tight arcing section, and then mount a fairly standard incline. But the Cabot Fishway has never worked right; Ted says the flow maybe was "too energetic." To compensate, structures were placed to reduce the current, but this may have compromised the attractiveness to the fish, many of which hold outside the fishway without entering it, and many that do try stop at the highly unnatural 180-degree turn. New modifications are planned, which may or may not work—engineers will fiddle—but the shad will decide.

As a researcher, Castro-Santos had examined the behavior of shad in the racing waters of the channel using radio transmitters placed in fish to get detailed signals on their movements. His findings demonstrate how the unnatural layout can defeat even determined individuals, and they offer little hope that small tweaks can compensate. Shad prefer to travel along the sides of the canal where the currents are weaker. The short ladder that actually gets shad over the dam is in the canal's east corner. But shad that swim up the fishway at the base of the dam enter the canal on its west side, and then would have to face the raging flow down the center of the canal to reach the ladder. To accommodate them, the actual entrance to the ladder in the east corner is on the west side, with the fish that enter it needing to make a swim up a lengthy additional stretch of ladder from there that runs parallel to the dam face. And shad that make it into the canal through the Cabot Fishway and then approach the final ascent on the east side are denied entry and must find their way to the west side, which, if they do, usually occurs by moving far back downstream away from the raging flow, before crossing over. Not only is the fishway entrance on the opposite side of the canal from the actual fishway, but it also is suspended as an orifice that extends only six feet beneath the surface, over depths of twenty feet.

If this is hard for the reader to follow, imagine how it is for the shad. Indeed, the tracking by Castro-Santos showed that shad have a difficult time sensing the attraction flow. Many burn their stored calories in this aquatic treadmill for only an hour or so; others go for as long as twelve hours before giving up and moving to quieter waters downstream to recuperate. It hits me that this setup is like asking the immense, wild herds of gazelle, zebra, and wildebeest of the Serengeti to migrate up a narrow man-made ramp, run miles through a long tunnel, and then find and squeeze their way through another ramp, all the while fighting a ferocious headwind. That any shad succeed at passing the Turners Falls Dam is a marvel to me.

We finish the tour at the other end of the canal, at a downstream passage chute just beyond the link with the Cabot Fishway. Spent adults and their young-of-the-year offspring have the option, if they find this outlet, of passing down to the main river through a concrete channel that resembles a scary water ride at an aquatic amusement park. Castro-Santos tells me that adults sometimes spend two weeks moving back and forth along the nearby electric-generation spillover area before committing to shoot through the narrow gap, trusting gravity to take them down to the real river. Squinting, I see hundreds of small dark shapes swimming in place in a broad triangle of flow that quickens before it launches down the flume. They are young shad, also stalled here—none are passing at this moment. I also notice a large bass holding deep in the gap, ready to rise up and snatch any that might dare to continue their migration.

Interlude II

A Shad's Journey, circa 2013

The shad breaks through the transparent capsule that had enveloped her, uncurls, and beats her tiny tail for the first time after having for ten days drifted as an egg near the water's surface. The shad is now little more than a delicate one-eighth-inch-long speck of life—albeit one with the potential to become a grand sojourner between river and ocean—but first she needs to survive her challenging earliest months. Her yolk sac provides a head start—three weeks' sustenance for growth and transformation to then become a predator of plankton.

The plankton in turn is supported by all the inorganic and organic materials loosened by agriculture and woodcutting and supplemented by human sewage discharges and washed by rains and the snowpack that melts from high in the watershed. This wealth of nutrients nourishes edible bacteria and also algae that bloom under the strengthening sun. The shad drifts with other larval shad, and with the eggs and larvae of other fishes in a small-scaled world; despite the depth and breadth of the river, their place in it is measured in inches. Because they can't swim far, the proportion that survives is closely linked to the richness of food in their own little patch; if either the timing or location of depletion of the yolk sac to the beginning of active feeding misses these localized spikes of production, survival is low. The shad encounters few older shad larvae, as the fitful spring weather and few early adult spawners have left the river bereft of older cohorts.

When her yolk sac is gone, the shad begins her existence in the middle of her food chain, needing both to eat and to avoid being eaten. But the density of plankton prey is barely sufficient because climate change has loosened the long-term relationships between the dual cues of sunlight and temperature in the timing of its production. She drifts with thousands of her kind as their numbers are winnowed by native perch, sunfish, and minnows, and a host of nonnative fishes as the survivors struggle to eat and grow.

One month later the shad has transformed into a small version of her adult form, graduating to larger prey such as insect larvae and gaining at least some capability to outmaneuver predators. By this time the shad's school is only of modest size. No longer mixed with their once-innumerable but now nearly extirpated relatives, river herring, the little shad are attacked relentlessly by larger fishes.

As the school moves downriver, the current slackens as they enter a large water-supply reservoir. The still waters offer little indication of flow direction, and the shad burn stored energy as they track diagonally back and forth in the mid-depths, with scant food in the offing, until they reach a giant wall. The school follows the barrier to one shore, reverses direction, and then yields to the subtle pull of water pouring over the top of the dam through a notch. They move close to the lip, hesitate, withdraw, and repeat the process several times over three hours until one individual bolts, drawing the others as if all are one.

It's an arduous trip down the flume as the bubbling flow crashes along the baffles that alternatingly line the sides of the sloping chute. They emerge, battered, only to face a battalion of jaws that ring the mouth of the apparatus. Most aggressive are the black bass, exotic fish with bellies already bulging but continuing to indulge in this easy feast; below them are eels that, despite their small mouths, manage to snatch some of the still-disoriented shad; and behind them is a rearguard of less-competitive native and nonnative predators alike that contribute to the carnage. The shad races high in the water, alarmed and confused, but instinctively follows the flow. From alongside a boulder near the tail of the pool, a rock bass rises to ambush her, but narrowly misses, and she scoots downstream.

The now severely diminished school settles into the river's estuary where it must endure murky waters contaminated with the effluents of the large city that envelops its shores. The shad seek the densest patches of prey, feeding day and night while attempting to avoid packs of juvenile bluefish and yearling striped bass that rapidly decimate their ranks. But some of their usual habitats become off limits. A severe August heat wave and the sudden discharge of human wastes caused by the failure of a municipal sewage treatment plant combine to drive down oxygen levels and to concentrate many of the estuary's fishes to a fraction of their normal range, which increases predation on the shad even further.

As autumn approaches, the rapidly growing shad turn downriver, making the necessary internal physiological adjustments to full-strength seawater to begin their marine existence. The shad's school enters the ocean from their natal river along with shad from some other Atlantic coast rivers, combining into larger schools. But a Murderer's Row of predators never ceases their attritions. Over the next several years she and her schoolmates move north in spring and south in autumn along the Atlantic coast, adding considerable length and weight. The school is steadily reduced in number, though, as the predators become larger, to include sharks, seals, dolphins, porpoises, and small whales. And nets take their toll—the shad mingle on the feeding grounds with far greater concentrations of sea herring. Industrial-size seines targeting sea herring incidentally sweep up many shad in the process of combing the waters for sardines.

As the remaining shad in the school move along the southern portion of their marine migratory path, some of the females become ripe with hundreds of thousands of eggs, while also sensing a pull to return to home waters. On this circuit, while swimming northward in spring, many small contingents break off from the main body of shad and stream toward their natal rivers. As the shad approaches the latitude of her river, she detects a faint but familiar odor—the unique mix of scents leaching from the bedrock, soils, organic matter, and man-made pollutants that characterize the watershed of her birth. This signal cues her to join thousands of her cohorts from her school and from others, swimming toward shore while homing to the increasingly intensifying smell of the river. The shad now consumes her last prey during the entire spawning run.

At the river's mouth the school hangs in the flow, adjusting osmotically to the low-salinity waters. Then they begin their journey toward the spawning grounds, pausing here and there to conserve energy in slower-moving river reaches while the shad's eggs continue to ripen and her abdomen swells. As they press farther upriver, the river narrows and becomes less deep. Progress is slow here as climate change has yielded little snowmelt to swell the river and the shad move cautiously through the clear shallows. A portion of her school moves along the shore of a state park and a whoop is heard as an angler wading in the water while walking on a row of curiously straight submerged boulders nets his second shad of the morning.

Driving through the fast water above the head of this reach, the shad's school enters a deep, slow-moving pool where many shad are already milling about. Exploring its confines, the cluster skirts a vertical wall. From a deep hole near the face of the dam, a large nonnative fish, a northern pike, rises to seize one of the smaller male shad, swimming off with its prize held sideways in its jaws, then settles back into its lair to turn the fish around headfirst to swallow it.

The shad in this reach are unsettled as the insistent migratory drive is stifled by the absence of an apparent path upriver and they circle and mill about for a few days. But there is a curious aquatic rumble from one corner of the pool where there is a tongue of current just strong enough to continually interest at least some of the shad. The school nears this feature, and with a burst of speed, three males enter the first chamber of the fish ladder and the remainder, including the shad, follows. The water fairly boils, but there is a dominant falling surge, and the vanguard senses it and instinctually plunges ahead and upward. Several shad bolt into the current, beating their tails rapidly, and then they pause on the calmer side of the next compartment, with the others following in sequence in what resembles an endurance contest. The school repeats these movements, slowly ascending some fifty vertical feet, before abruptly emerging into the quiet profundity of the reservoir. There they rest and mill about, before regrouping. The signal of flow direction in the big water body is faint and the school's movement toward it meanders, but two days later they enter the riverine section and drive upriver.

The cumulative delays in reaching the spawning reach leaves the shad forced to procreate in the warm, low-water conditions that characterize the end of the viable reproductive season. Dusk fades to darkness and the shad begin to stir, but because they are few, spawning occurs in fits and starts. A male repeatedly nudges the shad and they pair, each body quivers, and many of their sperm and eggs unite, but not with the seed of other shad given their paucity in this stretch of river. After resting the following day she repeats the performance that night.

Wearied and drained of eggs, the shad and a few other spawned-out fish regroup and begin to ride back downriver. The shad let the current assist them until they reenter the reservoir. A combination of a faint detection of flow direction, some meandering in their swim path, and cues realized from their recent reverse passage carries them to the great concrete wall. The shad's school wanders along it ceaselessly from shore to shore, seeking an exit. But the fair weather has decreased runoff, with little water spilling over the dam. More and more spawned-out shad arrive from upriver, adding to the frustrated assemblage. Few prey are encountered in the deep, still waters, and the shad steadily draw down their already-depleted energy reserves. After eight days of fruitless searching for a way out, the shad makes one more pass along the face of the dam and then quivers one last time, before sinking into the silt below.

Chapter 19

Favorable Currents, Fortunate Confluences

A river is more than an amenity, it is a treasure.
                    —*Oliver Wendell Holmes (1931)*

Steel-gray waves crash over the rock jetties as the forty-two-foot F/V *Dana Christine* plows through Delaware's Indian River Inlet. It's just past dawn on this April 2012 morning, and we have a lot of work to do. Captain Kevin Wark pilots the vessel south toward some depressions in the ocean bottom where gill netting had shown there were concentrations of Atlantic sturgeon, specimens traveling what I call the "sturgeon highway."

Captain Wark is a professional gillnetter from New Jersey who was hired on a grant by a young sturgeon researcher at Delaware State University, Dewayne Fox, to capture sturgeon as part of an ambitious look at the fish's migratory habits and a host of other biological characteristics. Captain Wark is a highly appropriate choice for this job; for a professional commercial fisherman, I find him unusually passionate about the importance of research and personally curious about what the team is learning. He also is an articulate voice in the Mid-Atlantic for the future of responsible fishing. Though only middle-aged, he's made an excellent living and will sooner than later retire to Maine, but he wants to leave the management of the industry in better shape than when he entered it. Also on board are Wark's mate, Mike Lohr, a quiet professional with a wry sense of humor, and Fox's master's student, Matt Breece. On the way to the grounds, we discuss the project. Breece says sturgeon are "way more prevalent now" than when they first began (their total catch now at the beginning of their fourth season reaching 473), which seems like an ironic statement given that the Atlantic sturgeon of the Mid-Atlantic have just been added—controversially, given their apparent abundance—to the Endangered Species List.

I can't help myself, so I run down a list of every species I suspect Wark fishes for, and he regales me with an ichthyologist's wet dream of what he's seen and caught over the years, their habits and trends, and special days like the time he and Lohr landed thirty-one large thresher sharks, each in the 200-pound class, as the marauders charged through schools of menhaden. But Wark is best known for having developed the lucrative coastal fishery for monkfish off New Jersey. Monkfish, also known as anglerfish, are wide-mouthed monsters that wiggle antennae to attract prey near their jaws as they sit on the bottom, but these light-boned predators also swim near the surface where they engulf marine birds as they dive underwater.

For decades the ugly monkfish were considered "trash fish," but as society discovers that the notion of fish as rubbish is a luxury—if not an insult—it's finding that all fish are edible, and a select few are truly delicious, such as monkfish. The problem with this fishery, however, was not so much the effect on monkfish, but that many Atlantic sturgeon were caught in the gill nets simultaneously as "bycatch." For a while these sturgeon could be harvested in the New York Bight, and they were, in suddenly large numbers, increasing from 13,000 pounds in 1988 to more than 220,000 pounds in 1990 in New Jersey alone. But as concern for the species grew, their harvest was prohibited there and everywhere along the US Atlantic coast by the Atlantic States Marine Fisheries Commission (ASMFC), with a landmark moratorium that could last up to forty years—enough time for two full generations of the slow-maturing fish to occur. But that in itself did not take all the pressure off the sturgeon, because some would perish in the gill nets set for monkfish. Concern grew at ASMFC and at NGOs such as the Natural Resources Defense Council, which began to push for endangered species status. More research was needed, and Dewayne Fox was one of a number of investigators to fill the void.

We are about to sink two giant gill nets—both 2,000-foot walls of either twelve- or thirteen-inch-wide mesh fished perpendicular to the beach in six fathoms of water. I can't wait to see the haul because recent catches were outstanding; over the two preceding days, the team landed forty-one sturgeon to more than 200 pounds from the sturgeon highway, plus a thresher shark.

What is the sturgeon highway? Atlantic sturgeon leave their natal rivers sometime over the first few years of life to mature in the richer waters of estuaries and the sea. And because they have both an extended adolescence that may last more than a decade before they spawn, and mature individuals, particularly females, only spawn every several years, Atlantic sturgeon are free to wander broadly in marine waters. And wander they do. But it is only recently that a wealth of details on these movements has begun to emerge. For most of the twentieth century, because sturgeon were both valuable and hard to catch, few individuals were marked with simple external tags—meaning, if a specimen was recaptured, knowledge would be gained from the geography of the release and recapture locations—minimal but still-useful information. The absence of a large mark-recapture database even left open the question of whether Atlantic sturgeon home to their natal rivers; it wasn't till Isaac Wirgin and I applied modern molecular approaches to these fish that homing could be inferred unequivocally through genetic signals.

But we are now leaving fewer secrets in the sea as concerns our important anadromous fishes with new tools being applied to sturgeon, striped bass, salmon, and other species. Chief among them are strings of radio-signal receivers being anchored in strategic locations along the ocean coast, in estuaries, and up rivers. More and more fish are being fitted with the corresponding radio transmitters, which generate a uniquely recognizable signal. Moreover, there is uniformity with all this equipment being furnished by one maker, so that receivers placed primarily for striped bass, for instance, will detect a sturgeon, and vice versa. Each year more are added; there are thirty in Delaware Bay and twenty-five in the ocean off Delaware, and more than a thousand along the rest of the coast. Couple this with molecular advances that now allow identification to river of origin with good confidence for most individuals, and today we are in an entirely new realm of knowledge acquisition, on the edge of an explosion.

After setting nets, Breece decides he wants to download data from the receivers, nine of them stretched in an array from a half-mile from shore out to ten miles. It takes a few minutes of scanning the waters to locate a flagged buoy, but soon Lohr hauls up the first receiver, a small unit that costs about

$1,500. Breece holds a laptop computer near the device, hits a couple of keys, and seconds later we know that the receiver has made 1,390 "detections" in the month since it was last downloaded. That's not 1,390 individual fish—many are multiple hits—but the computer program instantly processes these data, and they show that seventy-seven individuals passed the receiver, and that one of the sturgeon was later detected in the Delaware River and another in the Hudson.

Our hauls of the gill nets don't live up to the promise of the days before. On the first one, we catch only two winter skates and a clearnose skate. Wark says, "They're not moving yet," which is a critical point, because unlike a trawl net actively towed by a vessel, a fish must swim into a stationary gill net. If the fish are lolling around on the bottom, few will be caught.

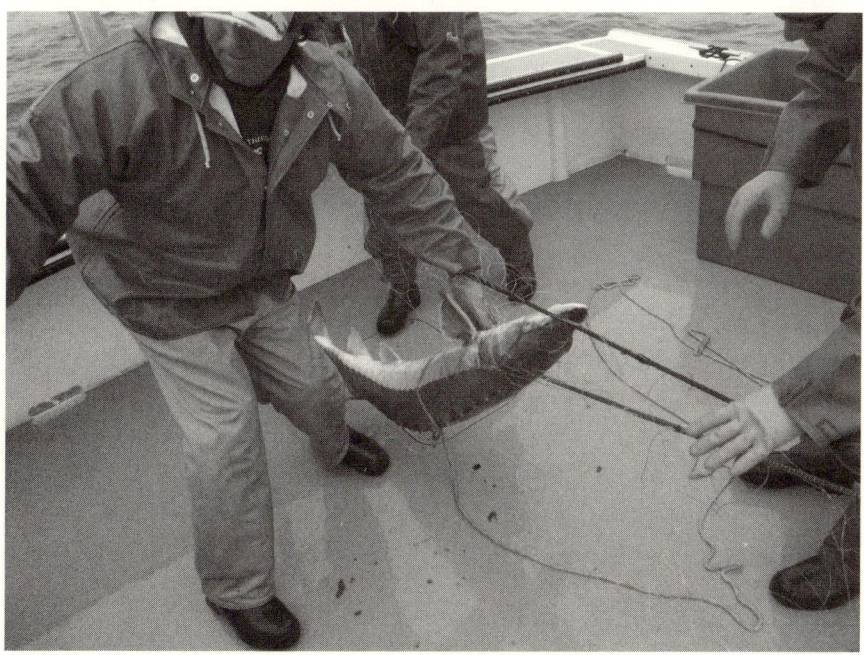

Captain Kevin Wark landing an Atlantic sturgeon caught in gill net for research, Delaware coast
John Waldman

The next net is cranked back out of what the team has named the "Helmet Hole." *Helmet* is commercial-fisherman speak for a horseshoe crab with its headdress-like shell, and the Helmet Hole often yields a mother lode. But again, little is moving down below; we land a skate, a monkfish, and a few helmets.

Wark is confident that as the day progresses, the sturgeon will start moving. We break the ice on the third haul, coming up with a fifty-pound Atlantic sturgeon, plus skates, a few helmets, and a seahorse. With a sturgeon in the holding tank, Breece gets to work. The specimen is lightly sedated with clove oil, weighed, and measured; next, a fin clip is made and placed in buffer for genetics work, and a passive integrated transponder (PIT) tag is embedded alongside a fin to provide a virtually permanent identifier to accompany the traditional external tag that in time will almost certainly fall off. Then Breece makes an incision in the fish's belly, slips in the little $350 transmitter, applies some anti-infection compound, and then sutures the wound. When the sturgeon seems fully recovered, we slip it back into the sea.

John Boreman's scheme—to make the modest Chesapeake Bay striped bass cohort of 1982 have the impact of a truly dominant year class—played out over the 1980s. Females of that vintage began to contribute to the spawning stock in 1986, and by 1987 females from 1982 and the subsequent two years began to dominate reproduction as the last of the big spawners from the glory days died off. With the new protections, the age structure of the spawning stock broadened in the 1990s, and the ever-growing females came back laden with even more eggs. The stage was set for a major rise, and after two poor spawning years in 1987 and 1988, it appeared to happen in 1989—though not without controversy. The 1989 year class–index value for Maryland waters of Chesapeake Bay was high enough to exceed an agreed-upon "trigger" in reopening the fishery that was based on the indexes' three-year running average, despite the previous two poor years—a situation worsened by the high 1989 index being heavily influenced by one exceptionally large seine haul of young stripers. Regardless of these misgivings, it was politically untenable not to follow through, and in 1990 the fishery was reopened, albeit tightly controlled.

The burgeoning striped bass populations of the Chesapeake and also the Hudson (enjoying the protection aimed at the Chesapeake) generated some explosive year classes in the 1990s, and in 1995 the ASMFC declared the Chesapeake Bay stock fully recovered. Indeed, stripers became so abundant that many fishermen and some researchers began to believe they had been restored too far—that their numbers relative to their forage foods were severely out of balance. I can recall taking my boat out fishing on western Long Island Sound in November in the early 2000s and seeing flocks of gulls hovering over schools of migrating striped bass that stretched as far as I could see in both directions as the gamefish drove baitfish to the surface in feeding frenzies; I'd guess they contained tens of thousands of stripers. This phenomenon essentially stopped in 2009 as a number of weak year classes were experienced, and fishermen began to speak again of an impending crash. Then the fourth-largest Chesapeake year class materialized in 2011. It appears that given the vagaries of nature and of balancing fishing harvests from many fisheries along the Mid-Atlantic and New England states, large swings in abundance will occur, but the painful lessons of the last great decline and the new, more precautionary measures put in place may avoid a repeat.

A little-discussed societal benefit of the striped bass crash is how it changed the mind-set of many sportfishermen. Before the strict regulations of 1982, the vast majority of striped bass landed by anglers were kept. From 1982 through the declaration of recovery, only small proportions were retained, but many were caught and released for sport. Anglers had learned that catch-and-release could be satisfying, especially when linked to a conservation cause. Unfortunately, the term *keeper* lingers in the hook-and-line fisherman's lexicon. *Keeper* implies almost preordained intent; a better, more neutral term is needed, such as *legal-size,* to help keep this noble fish ruling inshore Atlantic waters.

Even though it's only mid-March 2012, retired New York State fishery biologist Byron Young invites me to come out to eastern Long Island—the alewives are already running strong. In fact, a freakishly warm winter had some entering East End streams in February, an unusual occurrence. Young, a genial gentleman with a thick mustache, has a passion for alewife restoration, and keeping watch over various projects occupies much of his spring months. He's first going to show me a "healthy" run—something rare in New York waters.

We stop at a little bridge over Alewife Creek, the outlet to Big Fresh Pond, a natural eighty-five-acre "pothole" lake, a place where a huge chunk of ice melted after the last glacier receded. Two alewives are visible in the pool below the culvert under the bridge; one beats its tail in a rush and makes it over the lip of the culvert, holding in the darkness below the road. I know this to be a system that supports a major alewife run, as many as 70,000 each spring, but the scene is underwhelming. The creek is so small I could lie across it and block it, and it's only inches deep; being daylight, few fish are running. But there is another factor at play here: The alewives can enter the lake virtually unimpeded—there are no dams on the stream, only two minor lips on culverts. The fish come and go at will and sustain a massive run because of the sheer health of the system. It even reminds me of an Alaskan salmon run, albeit in drastically reduced form, as I spot the bodies of alewives claimed by raccoons, their highly nutritive brains consumed, just as done by Alaskan brown bears.

Next we visit a property on Big Fresh Pond owned by a friend of Young's, Dieter von Lehsten. Von Lehsten moved there at Christmas 1989, and when he sat on his floating deck the following March, he wondered why there was wild splashing going on around him. Intrigued, he later witnessed the little creek itself running silver, clogged from bank to bank with the bodies of the spawned-out adults returning to the sea. Summer brought ribbons of their young in schools along the shore as they matured in the lake. As often occurs when someone is exposed to such plenitude, an alewife aficionado was born. Now a passionate defender active in restoration, he soon became engaged in helping to maintain one of the best runs on Long Island. This includes helping to instate a ban of all fishing of alewives on Alewife Creek, a minor passage facility, to assist the spawning run during low-water conditions, and annual spring cleanup of the creek by the town prior to the run. He's also been impressed with the organismic cuing approximately two weeks prior to the alewives return, when ospreys, black-wing seagulls, cormorants, and various egrets congregate around the pond, eagerly awaiting the alewives to feast on them, as does von Lehsten himself, but only for psychic reasons.

We leave this nearly pristine gem of a system for downtown Riverhead, an old, once highly Polish center for East End farming now undergoing urban renewal. Right off the main traffic circle is Grangebel Park, a patch of green with a river flowing through it, the Peconic, a modest system by national standards but the Mighty Mississippi of Long Island. Fish passage has never been a priority of New York State, but Young's personal dedication caused him to team up with a similarly smitten high school teacher, Robert Conklin, and other enthusiasts, to follow a path that many successful grassroots restorations hew to.

It began with a "herring heave," a proof of concept where tens to hundreds of ready-to-spawn alewives were netted and released above the dam. Juveniles were seen, showing that the waters are amenable to reproduction. Next came a seasonal fish ladder, an aluminum contraption bolted to the dam in late winter and taken out before summer. Designed correctly and placed right, migrating alewives discovered it and passed over the dam, within a few years greatly enlarging the total run size. But then the ad hoc nature of the temporary ladder appeared unsatisfactory, and the righteousness of an enduring fix seemed apparent. Funds were found, engineers stepped in, the permanent ladder was built, the fish run grew exponentially, and the community took pride in its highly local but still meaningful act of ecological justice.

Today, though, the options for ladders go beyond metal and concrete. Young's group proposed a naturalized ladder, and with obvious pride he shows me the outcome. Though clearly man-made, it's not glaringly so, and the fish seem to perceive it the same way. From the pool below the dam, the alewives scoot up a straight rock-lined channel, climbing the 5-foot elevation along a 195-foot run. Partway up they must pass through a 10-foot-wide concrete shelf below a footbridge, but a few bolted-down boulders break up the flow enough to allow passage.

Next, Young shows me the results of the ladder; not far above the Grangebel Park Dam is a small trunk stream with a large pool below an earthen dam that lacks any passage provisions. The pool is running silver; alewives fill it from bank to bank, but this is because they are stalled here. A few break away and throw themselves up the flow along a rocky incline below a culvert, but it's hopeless. Scales scraped off the frustrated alewives wash down and collect on the bottom like little mirrors. Real progress has been made on the Peconic system—and there is momentum—but as almost everywhere else on the Eastern Seaboard, there is more work to be done.

Earth Day! What better way to celebrate seasonal renewal than to watch glass eels push inland from the sea? I wait by the lowermost bridge on Furnace Brook, the location where the little river meets brackish water pushing up with the rising tide from the Hudson. It's a classic late-April morning, and despite a chill breeze, birds are busy and singing as they prep for nesting; downstream, high above the marsh, a bald eagle makes repeated passes, perhaps looking for an early alewife. But just upstream I spot a strange contrivance: metal rebar stakes arranged in an arc, tipped with tennis balls to prevent canoe accidents, and about twelve feet of khaki-colored net in a U that arcs downstream and tapers to a tube staked farther up in the flow.

A car pulls up driven by a designated dad and out tumble two sleepy-looking teenage girls. Pam Brigleb and Amanda Bernstein get right to work, having performed these tasks for a month already as a school project that benefits them in multiple ways: They receive school credits, can add valuable lines on their résumés, and, most importantly, gain meaningful real-world experience with the most mysterious fish in the sea. The project, one of ten like it in the Hudson Estuary, benefits New York State, an entity like many others that sees the wisdom of having trained volunteers perform useful scientific monitoring for free. Win-win.

We wade out to the net's cod end and open it, and then the girls work their hands down the net's terminus, delicately plucking the fragile glass eels from the mesh. Slowly, the bucket bottom fills with the wiggling bodies of these lively, transparent miracles from the Gulf Stream. They seem like clones of each other, but then an elver appears, also diminutive—a transforming glass eel, now showing the greater size of the adult form.

Within ten minutes we have a final count: 54 glass eels and the elver. Bernstein says this is an average catch, but that on full or new moons, they can top 200. The promise of these eels is in their vigor and numbers, not in their size; too small to weigh alone in the field, Brigleb says twenty of them usually tally a bit over three grams. Glass eels strike me as being like the little seeds of a highly fecund plant seeking to germinate and take hold wherever fortune and their sheer vitality takes them. And a seed has also been planted in the approximately 275 other students involved in this hands-on learning; some will be inspired to pursue science, perhaps even a few as fish biologists. Maybe even Brigleb and Bernstein. Later they analyze glass eel numbers and water quality among Hudson tributaries, enter a poster showcasing their work in a national Intel competition, and win a Navy prize to London to further compete. There they stood out. Bernstein later said: "When we got to London, all the other projects were either robotics or medical technology . . . except for these eels!" There is something larger to this eel-monitoring project; maybe the hope for the freshwater-sea migratory fishes, like so much else, is with our youth. But how to reach them in the requisite numbers?

With such a knotty and dynamic subject as humans and diadromous fishes, all is not bad news—there have been some happy confluences. It's been announced that the Danskammer Generating Station, one of the four once-through Hudson River power plants, is going to shut down. Illegal fishing or "poaching" is beginning to be taken more seriously. In one instance in 2011, boaters reported thousands of dead striped bass floating in the ocean off Cape Hatteras, North Carolina. Commercial trawlers were "highgrading," i.e., replacing smaller fish already caught and already dead with larger ones so that they could return to port with the highest poundage for their catch quota. Word spread widely in the fishing press and outrage followed, with the Coast Guard citing several vessels for violating federal laws. And the unfortunate reality of forty-plus years of well-intentioned but expensive failure: In 2012 the US Fish & Wildlife Service announced that the Connecticut River Salmon Restoration project is over; the river's future should now focus on the far more viable shad restoration.

The pattern for recoveries of all sturgeons to be achingly slow was broken by shortnose sturgeon in the Hudson River. A tagging study showed there were about 15,000 of them in the 1970s. By the 1990s they'd increased fourfold, the most spectacular increase ever for a sturgeon worldwide. In fact, there may now be more shortnose sturgeon in the Hudson than in all its other dozen and a half populations combined. How it happened was a mix of protection by the Endangered Species Act and improved spawning habitat near Albany, New York, as a result of the Clean Water Act. And one more surprising factor; remember those exotic zebra mussels that altered the Hudson's ecology? One of their few positives is that shortnose sturgeon eat them, and so now have an endlessly available food source. Sturgeon in the United States and elsewhere also have had fishing pressure reduced on them by the development of sturgeon ranching in California—not in the environmentally questionable ways of salmon ranching in estuaries, but safely, far inland, in recirculating water systems.

More dams have come down. Among the largest barriers, a year before the Edwards Dam but to less publicity, the low-head Quaker Neck Dam on North Carolina's Neuse River was removed, opening 78 miles of mainstem habitat for American shad and striped bass, plus 900 miles of tributaries. Suddenly both species were being caught in Raleigh's city limits. This spurred a dam-removal groundswell on the Neuse system; one state biologist said, "There's tremendous potential; there's an increasing realization that this is fairly tangible, doable stuff." In 2004 the base of the Embrey Dam was destroyed, clearing the way for anadromous fish to return to 71 miles of the Rappahannock River in Virginia. Just a few weeks later a huge school of hickory shad passed upriver, so many that state biologists didn't need to keep sampling; one said they "could just see the fish everywhere." It took the collective efforts of the state fisheries management agency, a federal agency, an NGO devoted to the river, the city of Fredericksburg, and a Virginia senator on a campaign that began sixteen years earlier—but it got done.

And a radical solution to fish and dams is being tried on the Penobscot. In an unprecedented collaboration, the Penobscot Indian Nation, seven conservation groups, two hydropower companies, and state and federal agencies are working together to restore eleven species of sea-run fish to the Penobscot River, while maintaining energy production. The concept is an elegant compromise among these interests. Instead of deriving hydropower from dams on the main-stem Penobscot, those dams are being removed or fitted with fishways, and the hydropower turbines are being moved to or augmented on tributary streams farther up the valley. Migrations of salmon and alewives that swim deep into portions of the watershed will still be impeded, but opening the lower main stem means that sturgeon and striped bass will have access to 100 percent of their original habitat, and with other species will gain improved access to 1,000 miles of river, all with a net gain in electric output.

Fish passage and the broader goal of stream restoration are being pursued with varying strategies and degrees of passion among Atlantic states. As of 2013, the state government of New York still showed minimal interest, whereas Pennsylvania, Massachusetts, and Connecticut have performed heroic work. If I was to propose a *top-down* model for state governments to follow in pursuing stream restoration, I would choose Connecticut's. And to really make a difference, I'd take the supervisor of Connecticut's Inland Fisheries Division Diadromous Fish Program, Steve Gephard, and clone him, distributing the Gephards to any state that's not getting it done.

Gephard is a lean fifty-something with an anchorman's voice and a mix of affability, folksiness, and practical wisdom that can disarm or subdue anyone thwarting a dam-removal or fish-passage project he deems worth-while. He can be good cop, bad cop, or both at once, and he is equally at home working with hydrodynamic engineers or local politicians. Born near

Chicago but vacationing annually near East Haddam on the Connecticut River, as a boy fishing in a tributary, he snagged and reeled in a sea lamprey, thinking he'd performed a service by tossing the monster into the bushes. It was a formative experience; once he knew better, that act haunted him, and today he is one of the lamprey's few proactive advocates, actually seeding rivers with lamprey that have lost runs. Gephard is so interested in the success of sea lamprey that he organizes an annual nest survey in that same tributary, the Salmon River, shortly after the fish finish spawning.

One weekend morning I met him and a group of more than a dozen volunteers, many from his agency, at the first dam on the river. Gephard briefed the group on protocol, jumping into the shallows to point out some handy nests left by lamprey and showing us how to distinguish single nests from groupings.

Then Gephard and I broke off as a team to make counts on Fawn Brook, a second-order tributary to the Salmon, each of us taking from the middle of the flow to one of its banks. The lampreys had left plenty of gravel mounds behind, but some were already washing out, along with other natural gravel aggregations resembling nests. With Gephard checking my work, I soon felt confident in adding to the tally. My real problem was that walking against a current for hours in waters as high as waist-deep is exhausting; I felt as if I should have trained for it. Taking a lunch-break rest on the stream bank, I pumped Gephard on his program. To date, he and his team have completed about sixty fish-passage projects, no small number for what is the third-tiniest state. Gephard believes that "one project starts another." Education and gentle cajoling lay the groundwork for communities that may initially ask: "What is fish passage?" But after the first project is completed, the same community may come to him again and again, suggesting additional projects he hadn't envisioned.

Gephard's program is not funded internally at anywhere near a level where it can simply pay for anything but minor projects. But he does provide essential leadership; according to Gephard, "[The reality is that] our agency can't do all the work, but the work can't be done without our agency." A major task he takes on is to assign teams: to find funding, to work out technical details, and to explain and promote the project. And where some federal agencies seem to perform endless reviews and reprioritizations and feasibility studies before taking down a dam, Gephard's philosophy is to hover over the landscape watching for opportunities, and then to swoop in, keeping things simple, inexpensive—and even hands-on.

In Stonington the possibility emerged to build fish passage at the second dam on a small river, but that only made sense if the first dam could be passed. The first dam was a peculiar one: a thirty-foot-wide glacial erratic boulder that had wedged between rock outcrops, filled in on the sides with smaller boulders by early colonists. Because of its historical value, a modern fish ladder was not permitted, so Gephard suggested a stone-and-cement ladder that would look natural. Funds were scarce, so Gephard designed the ladder himself, rented a cement mixer, plugged into a neighbor's electrical outlet, and built it with staff from his project and The Nature Conservancy. In another project with The Nature Conservancy, a dam owner wanted a low dam on either side of an island taken out. Gephard avoided bureaucratic fuss and high costs by renting a jackhammer; the project only took as long as four days because the eager but naive biologists used the wrong tip on the tool.

After a long day counting the nests, Gephard invited the group back to his home for a barbecue, with staffers, interns, graduate students, and other volunteers talking fish passage—a terrific esprit de corps for what is a loose and highly effective commando team led by the master himself.

If I was to propose a *bottom-up* model for states to follow in pursuing stream restoration, I would choose the many informal and also the official Friends of the Blackstone. It's a classic cold-front morning as I wait for two heroes of the Blackstone River restoration project to meet me at the Slater Mill Dam. The river's waters have that typical autumn murk, but the slanting eastern sunlight gleams across the mill building, preserved as an integral part of America's legacy. Many see this location as the birthplace of the nation's Industrial Revolution. Before the late 1700s the spinning of cotton thread was a low-production cottage industry performed with hand-cranked equipment. An attempt in 1789 to power a mill at the falls failed, but by the next year Samuel Slater, an English immigrant with experience at a textile mill in England, had the experimental mill functioning. As I take in the scene, I imagine the mill some 200 years earlier. Well before first light many women workers, but also some men and children, would have streamed from their nearby homes to endure twelve-hour shifts from Monday through Saturday. To leave homes heated with wood or coal and lighted by tallow candles or whale oil, and then see gears and pulley belts driving machinery all across the factory floor courtesy of the ingenious use of the power of falling water, must have seemed as miraculous as splitting atoms for electricity does today.

John Marsland and Frank Geary greet me at the appointed time, and I quickly find they have the openness and relaxed good humor of men at peace with themselves, perhaps in part because they know they have made a difference. Marsland, a sprinkler fitter by trade, is barrel-chested and amicable; he defers to Geary, a wiry ex–pipe fitter with a sly sense of humor who knows how to size up and work people—qualities that likely made him a good shop steward. Both grew up along the middle reaches of the Blackstone River and watched it change for the better in their lifetimes. When Geary was younger he and his friends were told, "Don't go down by the river," because it was a toxic, foul place with visible dyes and foam. It was essentially nothing more than a conduit for the effluents of more than 100 mills that lined its banks. But in 1972 came the beneficial strictures of the Clean Water Act, and this was followed by the steady "deindustrialization" of America, which kindled both a cleaner Blackstone and hope among those who knew the river and saw it begin to improve.

Marsland is president of the Blackstone River Watershed Council/ Friends of the Blackstone, and Geary is head of its fish-passage project; as such, Geary is quick to open the rear of his car to show me framed engineer's drawings of the fish ladders to come. Fishways on three dams are planned, which will allow river herring to reach far inland, perhaps resulting in a run size numbering in the millions. Next we walk downstream of the mill, where Geary pulls out a creation he made that he uses when he speaks with school groups—a scale model of the dam, the planned fishway, and Indian Rock, the traditional Native American fishing outcrop, now half covered by the foundation of a building. Then he shows me the "book," a folder with all the relevant letters, maps, and blueprints that have been instrumental in

Model of fish ladder used to educate students,
shown by John Marsland at Slater Mill Dam on
Blackstone River, Pawtucket, Rhode Island
John Waldman

making the Blackstone restoration one of the most comprehensive in the nation. The man is clearly prepared, and I begin to see why the Friends get things done.

But there is much more to glean from them as we walk along the river. Their passion is evident, but now I sense their winning attitude. I am surprised that despite a minimal membership fee of only twenty dollars per year, and being based in the densely populated metropolis of Providence and Pawtucket, the Friends has little more than fifty dues-paying friends, and that the really active core group totals just a dozen or so. But because no one takes any salary, every project is lean and focused—all resources go directly to projects. And instead of having to conduct time-consuming, low-return fund-raisers to keep themselves moving forward, they've gone after large grants and they've received them, with about $4.5 million obtained so far.

Marsland started the organization in 1992 after an outside group was paid to begin a physical cleanup. He thought, "Why not just do it ourselves?" In a formative river makeover in 1972, 1,000 volunteers removed 10,000 tons of debris from its banks and waters. Geary had been aware of the river's rich history of salmon, shad, and river herring runs and wanted to see something done about restoring them. In June 1980, in recognition of the improving river conditions, the State of Rhode Island drafted an anadromous fishes restoration plan. But after the document sat on a shelf for nine years, Geary and the Friends grew impatient and decided they were going to make it happen. Geary rounded up a few allies to attend a kick-off meeting and then bluffed, allured, and cajoled the relevant agencies, telling each of them that everyone else was coming, this was going to be big and they're going to want to be a part of it, and if they don't, "they're going to go to hell." Everyone who should have been there showed—and they were on their way.

The best thing that can happen to a ravaged river is to gain national recognition that makes it the focus of restoration and preservation efforts and funding above the norm. Because of Slater's cotton mill and the central role the Blackstone played in the Industrial Revolution, in 1986 Congress recognized it as the John H. Chafee Blackstone River Valley National Heritage Corridor, coincidentally honoring the hero of striped bass restoration. Then, in 1998, President Clinton declared the Blackstone an American Heritage River, one of only fourteen in the nation. And in 2002 the US EPA and the US Army Corps of Engineers joined into a cooperative agreement to work on the Blackstone as an Urban River Restoration pilot project.

The Blackstone now enjoys an extraordinary degree of stewardship, a karmic balance for all its "hard work" in its last life. In addition to the Friends, the Blackstone River Watershed Association advocates for the river, educates about it, addresses harmful conditions, and organizes cleanups. It also works with the Blackstone River Valley National Heritage Corridor, an umbrella for a group of grassroots organizations that are trying to restore the river. The Fish Ladder Project alone lists seventeen local, regional, and national grassroots organizations, state and federal agencies, and local representatives and towns.

Work on the fish ladders was sent out to bid, but much has already been accomplished, and much more beyond passing fish will occur. Early on, the Friends pulled some 40,000 tires out of the river, enough to fill several eighteen-wheeler trucks that took them to recycling centers. The police asked if they might help get a vehicle connected with a crime out of the river. The Friends realized they could do a sunken-vehicle cleanup at the same time. A Friend in the auto-recycling business fronted $14,400 for a crane, and police divers went down to attach hooks and cables; five cars were brought up, with the insurance payments on them repaying the auto recycler. The auto recycler then turned over the insurance payments to the Friends. Green credits were found to allow the small hydroelectric facility at the Main Street Dam to shut down during fish-migration season. The fishways will have public viewing facilities. The Friends also intend to have automated fish counters and cameras in the fish ladders that will send real-time fish counts and videos to classrooms so that students can track the runs and relate passage to river conditions—a clear window into the ecology of what they should come to feel is *their* river.

In their own modest way, Marsland and Geary are a force of nature. I ask how they've managed to accomplish so much when many other rivers have piecemeal and lackluster efforts. Geary says, "We don't give up on anything." Marsland adds that they are widely known as the "can-do" organization. When they sit across from government agencies that they feel aren't coming through, Geary says they remind them that "you're forgetting one thing—we're the people—you're there to serve us." He adds, "We don't let them off the hook."

Geary also advocates unambiguous goals around which an appealing story can be told. Fish ladders fit that stratagem well; so does the broader aspiration. In the background of all this diffuse love and attention is a clear target that provides a focus: the Campaign for a Fishable/Swimmable Blackstone River by 2015.

The third haul on the *Dana Christine* was encouraging, so with raised expectations we make our fourth and final one. Perhaps it's a tide change or the time of day, but as Captain Wark predicted, the bottom of the sea seems to have come to life: Up come two black drumfish of 25 pounds or so, a smooth dogfish, some more skates and helmets, another 50-pound sturgeon, and then a six-foot-long sturgeon that weighs about 130 pounds. This last specimen is far from huge for a sturgeon, but it is impressive nonetheless. It will receive the same workup. But first I admire its patience as it rests in the tank, as if this momentary captivity means nothing to a descendant of an evolutionary line that stretches back to the lower Jurassic. The creature seems simultaneously "pokey" and magnificent, with its shovel-like snout, debonair barbels, odd underslung mouth, scutes borne like light armor, powerful caudal peduncle that drives its shark-like tail, and blood-red gills rhythmically pumping. We're going to send it on its way, and it may even report back to us from somewhere down the sturgeon highway, but I have the sense that this fish knows a timelessness humans will never share.

Chapter 20

Toward a New Stewardship

A fish is not an easy item to forge. . . . A fish is a truth.
—*Richard Flanigan,* Gould's Book of Fish

Along the Atlantic Seaboard we have so incrementally but also steadily rolled back the annual advancing armies of fish—as if we were fighting enemies instead of hindering friends—that we are hardly aware of our loss. But we have yielded more than many realize: ecosystems that once clicked like fine watches and now grind their gears as pale facsimiles of their former selves, a ready source of healthy protein, and our natural birthright to enjoy the wonders of a natural spectacle that causes the spirit to soar. We know our rivers are compromised; we can list litanies of individual environmental abuses. But if we want to understand the condition of any river with just one clarion signal of its health, we need only ask how its diadromous fishes are faring. Everything else becomes integrated with the answer to that single, simple question.

Daniel Pauly cautions of the "Aquacalypse"—the compounded effects of long-term cumulative overfishing, destructive fishing practices, and mismanagement of fisheries. This sorry assault has so harmed the ecological vigor of marine environments that Pauly and others have forecast "the end of fish." To know that so many diadromous fish stocks are down 90 percent or more from their peaks might be to agree with this prediction. But scores of them also persist, albeit at relict sizes, thus offering hope that there is still something tangible left to try to restore. To this end, I offer ten ideas that could make a difference.

Every dam should have an existentialist crisis. An existentialist crisis is when an individual questions the very foundations of his or her life; that is, whether this life has any meaning, purpose, or value. Since dams can't quite manage that level of introspection themselves, we should do it for them. The continued existence of every dam should be subjected to scrutiny. Most will quite apparently pass this screening with little review, yet a substantial portion will not be readily defensible. They will be shown to have endured to the present day largely through sheer inertia. We can't just wish dams away; it takes real effort. What constitutes the exact criteria for leaving a dam in place needs to be determined. Already many dams are being challenged on an ad hoc basis as to their continuing existence, not as a tidal wave but more as multiple ripples. It's not all that difficult to imagine a nationwide project or groundswell movement that includes a balance sheet for every dam that weighs societal benefit against environmental damage, including but certainly not limited to the ongoing harm they do to the survival of diadromous fishes. Those that fail to pass should be viewed not as benign and even beloved elements of the landscape, but as quietly latent environmental insults. The healing has begun; now it needs to accelerate.

Improve fish-passage technologies. Fish passage via ladders or elevators is the lesser evil versus no passage whatsoever. Fish passage has come a long way over the decades, although it remains an inexact science. It is a maturing discipline, with the advent of the Conte lab and other state-of-the-art research facilities, and with professional organizations that meet regularly to advance this cause. Not only should this momentum be encouraged, but modern fish passage also needs to be applied to many dams that lack such provisions. Special emphasis should be given to less-studied downstream passage, especially for adult fish that make it past dams to spawn. Ineffective returns downriver squander the potential of greater numbers and also larger sizes on future spawning runs.

An important and wide-open question is the long-term effects of passage on the fundamental characteristics of fish populations. For instance, it was found for one species that over forty years the nature of the ladder selected unnaturally against both the largest and smallest individuals. As for the responses of those laboratory *Menidia* to continued overfishing, could selection for a propensity of fish to climb a ladder come with a litany of other genetic effects?

Make hydroelectric licensing more responsive to short-term results. The Federal Energy Regulatory Commission (FERC) typically licenses hydroelectric plants for decades, under the supposition that these corporations require that period of certainty in their economic calculations. The problem is that, given the vagaries of designing effective fish-passage facilities, well-intentioned ladders and elevators that perform poorly are rarely overhauled to the necessary degree until the next license renewal.

Today many dams around the country licensed by FERC are under scrutiny. Hundreds of thirty- to fifty-year licenses FERC has issued to more than 2,000 privately owned hydroelectric dams are soon coming up for renewal. For instance, the thirty-year license for the Conowingo Dam expires in September 2014, with the owners seeking almost a half-century for the renewal. Under a 1986 change in the law, FERC cannot reinstate these licenses without subjecting dam owners to a rigorous standard of environmental accountability, and in situations of egregiously poor fish passage, there sometimes are ways to reopen a license. The ongoing failure of fish passage to work effectively in restoring runs past large main-stem dams suggests that some mechanism and funding for continued refinements during the license periods are needed, or that license durations should be shorter.

Tap nontraditional sources of hydroelectric power. The energy issue in the United States is terribly thorny; there are no easy answers to obtaining enough energy cleanly. Conservation is undoubtedly the lowest-hanging fruit, even if some dismiss it as "virtuous" (as if virtue is a vice). Beyond that, no energy source is totally free from environmental effects, and all have numerous trade-offs beyond them. To say hydropower dams should all be removed is glib and delusional. But if I was to offer a general guiding principle, it would be that the most environmentally egregious examples of each energy source should be put to rest, be it a windmill, a coal mine, or a hydro-dam.

One new approach to hydropower offers promise: the use of waters already claimed for municipal purposes, such as drinking water being delivered through pipes from storage reservoirs or wastewater treatment outfalls. Running these currents first through small turbines causes none of the ill environmental effects of a traditional installation on a dam. In San Diego a facility located on the water-distribution system generates 4.5 megawatts and $1.1 million in sales per year. Though projects like these are not likely to produce voluminous electricity, the future is likely to be far more of an energy mix of smaller contributors such as this than today's heavy reliance on fossil fuels, and perhaps will allow the dismantling of more ecologically harmful hydroelectric dams.

Fertilize roots. You may not have noticed it, but it's likely that a community group near you is trying to restore a stream or river or entire watershed. This is where essential passion and manpower often arise, but expertise may be lacking. Scientists should provide help freely, using their efforts to make a real difference, albeit at a small scale—acting locally—while also contributing to the more global science of river restoration. Institution of federal regulations such as the Clean Water Act was an absolutely essential top-down control to arrest environmental conditions spiraling out of control; the inherently optimistic and constructive endeavors of community-based river restoration is the sincerest form of bottom-up response to the opportunities this Act created.

Dam removals usually begin as true bottom-up grassroots initiatives. A desk-bound Washington bureaucrat doesn't one day jump up from his chair and shout, "The concrete monster must come down!" Grassroots institutions are critical to getting action on these removals and fish passage around much of the United States. Sometimes they coalesce organically and form regional coalitions, but on smaller systems they may not be well connected to their counterparts on other rivers. Better communication among grassroots organizations increases their efficiencies and allows them to face political obstacles more effectively. Funding these well-meaning groups with personal donations or foundation support is critical to their success.

Keep resource manager's feet warm. Many of the daily tasks of resource agency personnel can be thankless, and rarely are such underappreciated folks overpaid. But the cultures of some agencies sometimes evolve to acquire a sense of ownership of the resources they manage, including rivers and diadromous fishes. It's important that the public hold their feet to the fire if they remain unresponsive to worthwhile opportunities, much as Frank Geary did for the Blackstone when he reminded them who they actually work for.

Preserve "touchstone" rivers. Truly healthy Atlantic rivers are scarce these days, and undisturbed ones simply don't exist. Given the compromised to even ravaged states of many Atlantic rivers, those few that are relatively vigorous should be protected by all means, as something of an analogue to marine-protected areas. Though some larger rivers such as the Hudson and Potomac are listed as American Heritage Rivers, this is largely due to their places in history. Another designation is needed for those that somehow still function well ecologically, both as reservoirs of larger gene pools of fish and other river creatures, and as touchstones for comparisons with rivers being restored.

Among larger rivers these might include New Brunswick's Miramichi on the northern coast, an exceedingly healthy river that lacks any dams of consequence and still supports all of its diadromous species. On the Mid-Atlantic there is the Delaware, a perhaps surprising choice because of its legacy of industrial contamination, but the river has become cleaner, it lacks dams on its main stem, and it is enjoying recoveries of its anadromous fishes when most other large rivers are not. And to the south, I'd nominate Georgia's Altamaha, a big river with largely forested shores and a surprisingly low human population in its watershed. Then there are small systems, such as Long Island's Alewife Creek, a rare avatar of ecological health and productivity—another level of size worthy of protection.

Complexify rivers. In obliterating the original physiognomy of natural rivers with what Amy Roe calls the "occidental current," we also lost those other forms of shad and alewives seen in the 1800s that evolved with slightly different life histories from the dominant forms, differences likely connected to variations in habitat use and run timing. That different runs could exist within a single river was possible until then because fish were able to spread unimpeded throughout these watersheds, exploiting them fully. Those differences helped to provide greater abundances and more resilience by buffering poor reproduction of only one life-history form, and by maximizing use of the total river system.

Today, though, as Walter and Merritts showed, many rivers have been greatly simplified, with backwaters, channels, and islands lost, and sediments and bottom habitats altered, in addition to the broad impairment of dams cutting off access in the first place. Similar alternative life forms might eventually evolve again if we increase river qualities and quantities, and there are other kinds of resilience-promoting life-history variation within populations that fisheries biologists believe could be encouraged this way. The problem is how to accomplish this with rivers that are largely shackled in place. Ideally, we would set our rivers free to be what they want to be, as Laura Wildman recommends. More realistically, perhaps, river diversity could be engineered. Either way, the status quo model of the dam-blocked, semipermeable ditch is not tenable if we are to reverse these woefully shifted baselines.

Get a kid wet and educated. Communities near a river that supports diadromous fishes should get their kids wet, and more than once, in their school careers. Wet, as in donning a pair of waders and pulling a simple seine net and turning the ecology of a river from more than a nebulous textbook concept to the endearing reality of catching and handling its diverse life. Every time I've gone seining with students, their squeamishness fades as their excitement builds, with a longer-term curiosity taking hold that may ripple through the child's life in ways that benefit both the person and their sphere of influence.

This hands-on experience should be coupled with research projects on the history of the river, including every possible detail about their diadromous fish runs, generating a narrative that will simultaneously educate about the broad precolonial, agrarian, and industrial sagas of the nation, and about the particulars of their home waters and their own metaphorical backyards. To capture the attention of children is also to gainfully ensnare their parents, the information jumping across generations and perhaps feeding energy into local grassroots organizations.

We need more rats. River rats, precisely. *River rats* is a term of endearment for those souls who become addicted to the rhythms of life and the surprises so generously furnished by a river. The men who battled with Robert Boyle against the sins being committed or planned against the Hudson in the 1960s, and the community of people who became the hard-working Friends of the Blackstone three decades later, were mostly middle-aged and older folks who had spent formative years mucking around in their rivers. Unplanned, un-play-dated explorations through fishing, swimming, catching crayfish or crabs, building boats, skimming stones, and just poking around a river's shores and channels—even such highly compromised waterways as the Hudson and the Blackstone—led to lifelong love affairs, commitments returned later in life when time, resources, and the confidence that comes with age allowed payback for childhoods well spent. In a world where it seems we'll soon be required to wear a helmet just to walk to the store, there is much to be said for encouraging young people to reacquire the wisdom and sense of adventure that comes from just mucking around rivers.

Epilogue
Keep a Stiff Fin

In 2013 work began on a fish ladder for the Bronx River. In subsequent springs alewives should swim quietly through its murky South Bronx estuary till they butt against the dam, sense the ladder's attraction flow, work their way twenty-five feet up the first ramp, curve around a switchback, climb a twenty-foot section, followed by a ten-foot section, and then surmount the barrier—running silver to waters long closed to them, perhaps exorcising some ghosts along the way.

But there is so much more to be done to restore the great migratory fishes of Atlantic rivers. In "Saturday," the first chapter in *A Week on the Concord and Merrimack Rivers,* Thoreau expressed dismay at the future for shad, the collision between the "Corporation with its dam" and the hard-wired conduct of the fish, "with instinct not to be discouraged, not to be *reasoned* with, revisiting their old haunts, as if their stern fates would relent." Looking forward he envisioned them "armed only with innocence and a just cause. . . . Still wandering the sea in thy scaly armor to inquire humbly at the mouths of rivers if man has perchance left them free for thee to enter." He held hope that the losses he foresaw would resonate, stating, "It will not be forgotten by some memory that we were contemporaries."

Thoreau urged them on with this message: "Keep a stiff fin then, and stem all the tides thou mayst meet."

Acknowledgments

Much as a watercourse is the sum of its tributaries, in preparing this book, many individuals and institutions contributed to it along the way, both directly and indirectly. I thank my longtime collaborators and friends with whom I've worked on diadromous fishes and who have humored me in countless discussions about them, especially Karin Limburg and Ike Wirgin, and other colleagues and students too, including Dave Secor, John Boreman, Robert Boyle, Mary Fabrizio, Merry Camhi, George Jackman, and Mike Bednarski. Specific support for the book in myriad ways was also provided by Karen Alexander, Chris Bowser, Matt Breece, Jed Brown, Ted Castro-Santos, Adam Cook, Michael Dadswell, Dewayne Fox, Frank Geary, Steve Gephard, Carolyn Hall, Tom Horton, Adrian Jordaan, Ronald Kreisman, Tom Lake, Beth Lambert, William Leavenworth, Julie Maher, John Marsland, Karl Meyer, Chuck Nieder, Amy Roe, Daryl Pierce, Ted Steinberg, David Strayer, Dieter von Lehsten, Kevin Wark, Laura Wildman, Tim Wildman, Karen Wilson, Byron Young, and Joseph Zaientz. All or portions of the manuscript were read by Karin Limburg, Amy Roe, Jocelyn Courtney, Ted Castro-Santos, Karl Meyer, Ted Steinberg, and Carol, Laura, and Steve Waldman.

I remain grateful to the Hudson River Foundation for Science and Environmental Research for providing a supportive platform for two decades for investigations of the Mighty Hudson and diversions into the international realms of the conservation of sturgeons and shads, and to Queens College and the City University of New York for a later academic home—one that included the marvel of a yearlong sabbatical to focus on this project. Financial support at key times was provided by Joan Davidson and Furthermore: A Program of the J. M. Kaplan Fund, and by a Professional Staff Congress–City University of New York Research Award. And I will be everlastingly thankful to the Rockefeller Foundation for a full month of writing in undisturbed bliss in an Italian palace courtesy of a Bellagio Fellowship.

Good times angling for, and talking about, many of the species mentioned were shared with Rob Maass, Steve Sautner, Dave Taft, and Ike

Acknowledgments

Wirgin. I also owe appreciation to Paul Greenberg for his psychic support and for introducing me to my agent for this book, David McCormick. Janice Goldklang, executive director of editorial at Globe Pequot/Lyons Press deserves a special thank-you for believing in *Running Silver*. And project editor Meredith Dias ably put all the parts together.

Family members Josie, Ed, Shel, Sher, Vivian, Mike, Devin, Kyle, Tom, Conor, Jocelyn, and Chris were encouraging and supportive throughout. Finally, deepest gratitude to my wife, Carol, and my children, Laura and Steve, who have been the finest company on my journey down this lengthy river.

Endnotes

Preface

x. "Who hears the fishes when they cry?": Thoreau, Henry David. *A Week on the Concord and Merrimack Rivers* (self-published, 1849).

Prologue

xiii–xiv. Thoreau: Thoreau, *A Week on the Concord and Merrimack Rivers*.

xv. Edwards Dam removal: McPhee, John. *The Founding Fish* (New York: Farrar, Straus and Giroux, 2002).

xvii. Mianus River fishway: www.greenwichct.org/upload/media library/5b3/Fish_Mianus_Pond_Fishway.pdf. A live camera tracking the run through the fishway is available in spring at www.greenwich ct.org/Government/Departments/Conservation_Commission/ Mianus_Pond_Fishway_Underwater_Camera/.

xvii. Otoliths for estimating age of fish: Campana, S. E., and J. D. Neilson. "Microstructure of Fish Otoliths," *Canadian Journal of Fisheries and Aquatic Sciences* 42 (1985): 1014–32.

Chapter 1

2. "We are set down eighty miles . . .": Letter from the Council in Virginia to the Council in England 1607. Cited in Wharton, James, *The Bounty of the Chesapeake: Fishing in Colonial Virginia* (Charlottesville: University Press of Virginia, 1957).

2. "Yea, when a heape of stones . . .": Puritan account cited in Bakeless, John, *America as Seen by Its First Explorers: The Eyes of Discovery* (Mineola, NY: Dover, 2011).

2. "We had more sturgeon than could be devoured by dog or man . . .": Statement by John Smith in 1609. As quoted in Wharton, *The Bounty of the Chesapeake*.

2. "In the spring of the year, herrings . . .": Beverley, Robert. *The History and Present State of Virginia* (London, 1705). As quoted in Wharton, *The Bounty of the Chesapeake*.

2. "[T]he greate smelts passé up . . .": Account by colonist John Pory, as quoted in Bakeless, *America as Seen by Its First Explorers*.

2. "There are such multitudes, . . .": Morton, Thomas. *New English Canaan, or New Canaan, Containing an Abstract of New England* (Amsterdam, 1637), as quoted in Fearing, D. B., "Some Early Notes on Striped Bass," *Transactions of the American Fisheries Society* 32 (1903): 90–108.

3. "The sturgeons be all over the country, . . .": Wood, William. *New England's Prospect*. Originally published in 1634 (Amherst: University of Massachusetts Press, 1993).

3. "[D]uring one month the fish ascend . . .": From 1723 account of Sebastien Râle, a Jesuit missionary to the Abenaki, cited in Brumbach, H. J., "Anadromous Fish and Fishing: A Synthesis of Data from the Hudson River Drainage," *Man in the Northeast* 32 (1986): 35–66.

3. "When they spawn, all streams and waters . . .": Statement by William Byrd II. *The Natural History of Virginia* (1728). As quoted in Wharton, *The Bounty of the Chesapeake*.

3. Status and trends of the Atlantic's diadromous fishes: Limburg, Karin, and John Waldman. "Dramatic Declines in North Atlantic Diadromous Fishes," *Bioscience* 59 (2009): 955–65.

4. The International Union for the Conservation of Nature: www.iucn.org.

Chapter 2

8. anadromy and catadromy in relation to freshwater and marine productivity: Gross, M. R. "Evolution of Diadromy in Fishes," *American Fisheries Society Symposium* 1 (1987): 14–25.

10. Hawaiian gobies: Keith, P. "Biology and Ecology of Amphidromous Gobiidae of the Indo-Pacific and the Caribbean Regions," *Journal of Fish Biology* 63 (2003): 831–47.

11. theory on evolution of diadromy: Gross, *Evolution of Diadromy in Fishes.*

11. diadromy in relation to fish evolutionary trees: McDowall, R. M. "The Evolution of Diadromy in Fishes (Revisited) and Its Place in Phylogenetic Analysis," *Reviews in Fish Biology and Fisheries* 7 (1997): 443–62.

12. variation in Hudson River shad: Mansueti, R. and H. Kolb. "A Historical Review of the Shad Fisheries of North America," *Maryland Department of Research and Education Publication* 97 (1953): 293 pages.

13. variation in salmon: Buck, Richard. *Silver Swimmer: The Struggle for Survival of the Wild Atlantic Salmon* (New York: Lyons Press, 1993).

13–14. Life-history variation in shad in Saint John River: Carscadden, J. E., and W. C. Leggett. "Life History Variations in Populations of American Shad, *Alosa sapidissima* (Wilson), Spawning in Tributaries of the St John River, New Brunswick," *Journal of Fish Biology* 7 (1975): 595–609.

14. energy expenditures by spawning American shad: Glebe, B. D., and W. C. Leggett. "Latitudinal Differences in Energy Allocation and Use during the Freshwater Migrations of American Shad *(Alosa sapidissima)* and Their Life History Consequences," *Canadian Journal of Fisheries and Aquatic Sciences* 38 (1981): 806–20.

14. "science of stock identification has taken these fish apart": Waldman, J. R. "The Importance of Comparative Studies in Stock Analysis," *Fisheries Research* 43 (1999): 237–46.

14–15. stock identification as applied to striped bass: Waldman, J. R., J. Grossfield, and I. Wirgin. "Review of Stock Discrimination Techniques for Striped Bass," *North American Journal of Fisheries Management* 8 (1988): 410–25.

16. King of Fish: Montgomery, D. *King of Fish: The Thousand-Year Run of Salmon* (Boulder, CO: Westview Press, 2003).

17. alewives as nutrient sources in Connecticut: West, D. C., A. W. Walters, S. Gephard, and D. M. Post. "Nutrient Loading by Anadromous Alewife *(Alosa pseudoharengus)*: Contemporary Patterns and Predictions

for Restoration Efforts," *Canadian Journal of Fisheries and Aquatic Sciences* 67 (2010): 1211–20.

17. sea lamprey as a nutrients source: Nislow, Keith H., and Boyd E. Kynard. "The Role of Anadromous Sea Lamprey in Nutrient and Material Transport between Marine and Freshwater Environments," *American Fisheries Society Symposium* 69 (2009): 485–94.

Chapter 3

19. rainbow smelt historical distributions: Bernatchez, L., and S. Martin. "Mitochondrial DNA Diversity in Anadromous Rainbow Smelt (*Osmerus mordax Mitchill*): A Genetic Assessment of the Member-Vagrant Hypothesis," *Canadian Journal of Fisheries and Aquatic Sciences* 53 (1996): 424–33.

21. *The Run:* Hay, John. *The Run* (Garden City, NY: Doubleday, 1959).

22. *Four Fish:* Greenberg, Paul. *Four Fish: The Future of the Last Wild Food* (New York: Penguin, 2010).

23. life history of American shad. Limburg, K. E., K. A. Hattala, and A. Kahnle. "American Shad in Its Native Range," *American Fisheries Society Symposium* 35 (2003): 125–40.

25. shad commerce: McPhee, *The Founding Fish.*

26. fishing with lobster bait: Wood, *New England's Prospect.*

26–27. spearing striped bass from horseback: Letter written in 1875 by Colonel T. J. Randolph as quoted in Wharton, *The Bounty of the Chesapeake.*

28. surfcasting for striped bass: Matthiessen, Peter. *Men's Lives: The Surfmen and Baymen of the South Fork* (New York: Random House, 1986).

29. "They ascend falls by clinging to the stones": Thoreau, *A Week on the Concord and Merrimack Rivers.*

30. experimental evidence for non-homing by sea lamprey: Bergstedt, R. A., and J. G. Seelye. "Evidence for Lack of Homing by Sea Lampreys," *Transactions of the American Fisheries Society* 124 (1995): 235–39.

30. theory and genetic evidence for non-homing by sea lamprey: Waldman, J., C. Grunwald, and I. Wirgin. "Sea Lamprey *Petromyzon marinus:* An Exception to the Rule of Homing in Anadromous Fishes," *Biology Letters* 4 (2008): 659–62.

30. chemical communication in sea lamprey: Bjerselius, R., and eight coauthors. "Direct Behavioral Evidence that Unique Bile Acids Released by Larval Sea Lamprey (*Petromyzon marinus*) Function as a Migratory Pheromone," *Canadian Journal of Fisheries and Aquatic Sciences* 57 (2000): 557–69.

31. sea lamprey trawled at a depth: Collette, B. B., and G. Klein-Macphee. *Bigelow and Schroeder's Fishes of the Gulf of Maine* (Washington, DC: Smithsonian, 2002).

31. native status of sea lamprey in Lake Ontario: Waldman, J. R., C. Grunwald, N. K. Roy, and I. I. Wirgin. "Mitochondrial DNA Analysis Indicates Sea Lamprey (*Petromyzon marinus*) Indigenous to Lake Ontario," *Transactions of the American Fisheries Society* 133 (2004): 950–60.

32. world sturgeon diversity: Birstein, V. J., J. R. Waldman, and W. E. Bemis. *Sturgeon Biodiversity and Conservation* (Dordrecht: Kluwer, 1997).

32. sturgeon working like hogs through mud in Tampa Bay: *Bulletin of the US Fish Commission for 1888,* Vol. 8 (Washington, DC: Government Printing Office, 1890).

33. why sturgeon leap: Recent scientific findings may be found in Watanabe, Y., and eight coauthors. "Swimming Behavior in Relation to Buoyancy in an Open Swimbladder Fish, the Chinese Sturgeon," *Journal of Zoology* (2008): 1–10. Some earlier speculations are summarized in Waldman, John. "The Lofty Mystery of Why Sturgeon Leap," *New York Times,* October 21, 2001.

34. "a kind of philosopher among fishes": Carey, Richard Adams. *The Philosopher Fish: Sturgeon, Caviar, and the Geography of Desire* (New York: Counterpoint, 2005).

34. DNA analyses of sturgeon: Wirgin, I. I., J. E. Stabile, and J. R. Waldman. "Molecular Analysis in the Conservation of Sturgeons and Paddlefish," *Environmental Biology of Fishes* 48 (1997): 385–98.

35. Leif Ericson: Netboy, Anthony. *The Salmon: Their Fight for Survival* (Boston: Houghton Mifflin, 1973).

38. "*. . . the World's Most Mysterious Fish*": Prosek, James. *Eels: An Exploration from New Zealand to the Sargasso, of the World's Most Mysterious Fish* (New York: HarperCollins, 2010).

39–40. seminal paper on eel reproduction: Schmidt, Johannes. "The Breeding Places of the Eel," *Philosophical Transactions of the Royal Society of London, Series B, Containing Papers of a Biological Character* 211 (1923): 179–208.

40–41. eel species in Iceland: Avise, J. C., and five other authors. "The Evolutionary Genetic Status of Icelandic Eels," *Evolution* 44 (1990): 1254–62.

41. eels in Adirondack Mountains: Mather, Fred. *Memoranda Relating to Adirondack Fishes* (Albany, NY: Weed, Parsons and Company, 1896).

41–42. eels at Niagara Falls: Clinton, D. "Some Remarks on the Fishes of the Western Waters of the State of New York, in a Letter to S. L. Mitchell," *Transactions of the Literary Society of New York* 1 (1815): 493–501.

42. Sigmund Freud on eels: Freud, Sigmund. "Observations on the Form and Fine Structure of the Looped Organs of the Eel, Organs Considered as Testes," as cited in Prosek, *Eels.*

42. evolution on islands: Quammen, David. *The Song of the Dodo: Island Biogeography in an Age of Extinctions* (New York: Scribner, 1997).

44. history of brook trout on Long Island: Karas, Nick. *Brook Trout: A Thorough Look at North America's Great Native Trout—Its History, Biology, and Angling Possibilities* (New York: Lyons Press, 1997).

46. conservation of Atlantic whitefish: "Recovery Strategy for the Atlantic Whitefish (*Coregonus huntsmani*) in Canada," *Species at Risk Act Recovery Strategy Series* (Fisheries and Oceans Canada, Ottawa, 2006).

46–47. New England freshwater fishes and Northeastern Banks Refugium: Schmidt, R. E. "Zoogeography of the Northern Appalachians," pages 137–59 in C. H. Hocutt and E. O. Wiley, editors. *The Zoogeography of North American Freshwater Fishes* (New York: Wiley, 1986).

Chapter 4

50–51. stream order: Strahler, Arthur N. "Hypsometric (Area-Altitude) Analysis of Erosional Topology," *Geological Society of America Bulletin* 63 (1952): 1117–42.

51. River Continuum Concept: Vannote R. L., G. W. Minshall, K. W. Cummins, J. R. Sedell, C. E. Cushing. "The River Continuum Concept," *Canadian Journal of Fisheries and Aquatic Sciences* 37 (1980): 130–37.

51. "All land represents a downhill flow of nutrients from the hills to the sea": Leopold, Aldo. "Lakes in Relation to Terrestrial Life Patterns," pages 17 to 22 in *University of Wisconsin Symposium, Volume on Hydrology* (Madison, 1941).

52. Serial Discontinuity Concept: Ward J. V., and J. A. Stanford. "The Serial Discontinuity Concept of River Ecosystems." T. D. Fontaine, S. M. Bartell, editors. *Dynamics of Lotic Ecosystems* (Ann Arbor, MI: Science Publications, 1983).

52. spatial scales: Frissell, C. A., W. J. Liss, C. E. Warren, and M. D. Hurley. "A Hierarchical Framework for Stream Habitat Classification: Viewing Streams in a Watershed Context," *Environmental Management* 10 (1986): 199–214.

53–54. shape of modern rivers: Davies, N. S., and M. R. Gibling. "Cambrian to Devonian Evolution of Alluvial Systems: The Sedimentological Impact of the Earliest Land Plants," *Earth-Science Reviews* 98 (2010): 171–200.

54–56. legacy effects of milldams: Walter, R. C., and D. J. Merritts. "Natural Streams and the Legacy of Water-Powered Mills," *Science* 319 (2008): 299–304.

56–57. ecological effects of beavers: Naiman, R. J., C. A. Johnston, and J. C. Kelley. "Alteration of North American Streams by Beaver," *Bioscience* 38 (1988): 753–62.

57. beaver and "reverse-greenhouse effect": Varekamp, J. C. "The Historic Fur Trade and Climate Change," *Eos* 87 (2006): 593, 596–97.

Chapter 5

63–64. shifting baselines syndrome: Pauly, Daniel. "Anecdotes and the Shifting Baseline Syndrome of Fisheries," *Trends in Ecology and Evolution* 10 (1995): 430.

64–66. approaches to historical ecology: Lotze, H. K. and B. Worm. "Historical Baselines for Large Marine Animals," *Trends in Ecology and Evolution* 24 (2008): 254–62.

66. historical ecology reconstruction using sportfishing photographs: McClenachan, L. "Historical Declines of Goliath Grouper Populations in South Florida," *US Endangered Species Research* 7 (2009): 175–81.

66–67. oral history of Gulf of Maine cod fisheries: Ames, Edward P. "Atlantic Cod Structure in the Gulf of Maine," *Fisheries* 29 (2004): 10–28.

71. "ecosocial anomie": Limburg and Waldman, *Dramatic Declines in North Atlantic Diadromous Fishes.*

Chapter 6

72. early interactions between Indians and colonists: Mann, Charles. "Native Intelligence," *Smithsonian* 36 (December 2005).

73. numbers of Native Americans in New England: Cronon, William. *Changes in the Land: Indians, Colonists, and the Ecology of New England* (New York: Hill and Wang, 1983).

73. Paterson weir: Lutins, Allen. *Prehistoric Fish Weirs in Eastern North America.* Master's Thesis, University of New York, Binghamton, 1992.

74. Spearfish Moon: McPhee, *The Founding Fish.*

76–77. Indian fishing methods: Buckley, B. and S. W. Nixon. "An Historical Assessment of Anadromous Fish in the Blackstone River," *Final Report to the Narragansett Bay Estuary Program, the Blackstone River Valley National Heritage Corridor Commission, and Trout Unlimited* (Narragansett: University of Rhode Island, 2011).

77. Indian way of catching sturgeon: Beverly, "The History and Present State of Virginia, 1705." As quoted in Wharton, *The Bounty of the Chesapeake.*

77. petroglyphs along Susquehanna River: Lenik, Edward J. *Making Pictures in Stone: American Indian Rock Art of the Northeast* (Tuscaloosa: University of Alabama Press, 2009).

78–79. importance of diadromous fish to Indian diets: Schindler, Bill. "Rethinking Middle Woodland Settlement and Subsistence Patterns in the Middle and Lower Delaware Valley," *North American Archaeologist* 29 (2008): 1–12.

79–80. importance of eels to Native Americans: Casselman, J. M. "Dynamics of Resources of the American Eel (*Anguilla rostrata*): Declining Abundance in the 1990s," chapter 18 in *Eel Biology* (Tokyo: Springer-Verlag, 2003).

80–81. American sturgeon colonizing Baltic Sea: Ludwig, A., and eight coauthors. "When the American Sea Sturgeon Swam East," *Nature* 419 (2002): 447–48.

81. American sturgeon colonizing French waters: Desse-Berset, Nathalie. "First Archaeozoological Identification of Atlantic Sturgeon (*Acipenser oxyrinchus Mitchill*), 1815, in France," *Comptes Rendus Palevol* 8 (2009): 717–24.

82. decline of Native Americans following European contact: Mann, *Native Intelligence.*

Chapter 7

83. Coonamessett Herring War: Wildman, Laura. "Dam Removal: A History of Decision Points," pages 1–10 in *The Challenges of Dam Removal*

and River Restoration. De Graff, J. V., and J. E. Evans, editors. *Reviews in Engineering Geology* XXI, 2013.

83–84. Native American organization on lands: Steinberg, Theodore. *Nature Incorporated: Industrialization and the Waters of New England* (New York: Cambridge University Press, 1991).

84. shad reaching settlers near Lake Otsego: Cooper, John Fenimore. *The Pioneers: The Sources of the Susquehanna; a Descriptive Tale* (1823) (Cambridge, MA: Belknap Press of Harvard University Press; Reprint edition, 2011).

86. European fishing history: Roberts, Callum. *The Unnatural History of the Sea* (Washington, DC: Island Press, 2007).

86. exploratory trip and improvised fishing in Chesapeake Bay: John Smith in 1608, as quoted in Franklin, H. Bruce, *The Most Important Fish in the Sea* (Washington, DC: Island Press, 2007).

86–87. Regulations concerning sturgeon fishing: Strachey, William. "Article, Lawes and Orders for the Colony in Virginia Brittania, 1612," as quoted in Pearson, John C., "The Fish and Fisheries of Colonial Virginia," *The William and Mary Quarterly* 23 (1943): 278–84.

87. Divine Providence and Taunton River: Emery, Samuel Hopkins. *History of Taunton, Massachusetts: From Its Settlement to the Present Time* (Syracuse, NY: D. Mason and Co., 1893).

87. Kennebec River fisheries: Letter by Sebastien Râle, 1723, as quoted in Allen, William, *The History of Norridgewock* (Norridgewock, ME: Edward J. Peet, 1849).

87–88. smelt fishing on the Restigouche River: Rowan, John, J. *The Emigrant and Sportsman in Canada* (Montreal: Dawson Brothers, 1881).

88. lack of salt to preserve fish: Wharton, *The Bounty of the Chesapeake.*

88. leftover shad as fertilizer: Roe, Amy W. *Swimming the Occidental Current: A Resource Hermeneutics of Fishery Collapse.* Doctoral Dissertation, University of Delaware, 2011.

89. Potomac River archaeological fish remains: Miller, H. M. "Transforming a 'Splendid and Delightsome Land': Colonists and Ecological Change in the Chesapeake 1607–1820," *Journal of the Washington Academy of Sciences* 76 (1986): 173–87.

89. "widow's hauls" of shad: Gerstell, Richard. *American Shad in the Susquehanna River Basin: A Three-Hundred-Year History* (University Park: University of Pennsylvania Press, 1998).

89–90. Stump Farm fishing operation and giant catch: Willis, H. "Shad Fisheries of the Susquehanna River Fifty-Six Years Ago," *Bulletin of the United States Fish Commission for 1881* (Washington, DC: Government Printing Office, 1882).

90. "shad batteries": Gerstell, *American Shad in the Susquehanna River Basin*.

91. eels in St. Lawrence River and Lake Ontario: Goode, G. B. "The Eel Question," *Transactions of the American Fisheries Society* 10 (1881): 81–124.

92. anadromous fish as food source during Revolutionary War: Watts, Douglas. *Alewife: A Documentary History of the Alewife in Maine and Massachusetts* (Augusta, ME: Poquanticut Press, 2012).

92–93. Atkinson's Mill Dam Raid and Great Safe Harbor Shad War: Gerstell, *American Shad in the Susquehanna River Basin*.

93. milldams on Cobbosseecontee Stream: Watts, *Alewife: A Documentary History of the Alewife in Maine and Massachusetts*.

95. "Shad, at one time, entered every river": Norris, Thaddeus. *American Fish-Culture, Embracing all the Details of Artificial Breeding and Rearing of Trout; the Culture of Shad, Salmon, and other Fishes* (Philadelphia: Porter & Coates, 1868).

Chapter 8

96. dissertation on history of Susquehanna River shad: Roe, *Swimming the Occidental Current*.

98. "one of the most furious, perilous, and ungovernable torrents in the world": Maryland Commissioners. *Report of the Maryland Commissioners on*

a Proposed Canal from Baltimore to Conewago (Baltimore: Fielding Lucas, Junior, 1823).

99. Webb Fishery on Susquehanna: *Bulletin of the United States Fish Commission, Vol. 1, for 1881* (Washington, DC: Government Printing Office, 1892).

99–101. commodification and dam-building history of Susquehanna River: Roe, *Swimming the Occidental Current.*

102. "America's hardest-working river": Klyberg, Albert. "Mr. Blackstone's River," *Narragansett Bay Journal* (Spring 2012): 1–4.

102–4. history of Blackstone River colonial fisheries: Buckley and Nixon, *An Historical Assessment of Anadromous Fish in the Blackstone River.*

Chapter 9

108. overview of PCB contamination in Hudson River: "PCB Contamination of the Hudson River Ecosystem: Compilation of Contamination Data Through 2008." Hudson River Natural Resource Trustees, National Oceanic and Atmospheric Administration, Silver Spring, Maryland, 2013.

108–9. "temperature-oxygen" squeeze: Niklitschek, E. J., and D. H. Secor. "Modeling Spatial and Temporal Variation of Suitable Nursery Habitats for Atlantic Sturgeon in the Chesapeake Bay," *Estuarine, Coastal and Shelf Science* 64 (2005): 135–48.

109. ship strikes of sturgeon: Brown, J. J., and G. W. Murphy. "Atlantic Sturgeon Vessel-Strike Mortalities in the Delaware Estuary," *Fisheries* 35 (2010): 72–83.

110–11. Hudson River battles against power plants: Boyle, Robert H. *The Hudson River: A Natural and Unnatural History* (New York: Norton, 1969).

111. Data on fish mortalities caused by Hudson River power plants: Reports on specific studies are produced by consultants to the utility companies; most are available at the library of the Hudson River Foundation for Science and Environmental Research, New York, New York.

112. "compensation" in fish populations: Rose, K. A., J. H. Cowan Jr., K. O. Winemiller, R. A. Myers, and R. Hilborn. "Compensatory Density Dependence in Fish Populations: Importance, Controversy, Understanding and Prognosis," *Fish and Fisheries* 2 (2002): 293–327.

112. retrospective overview on Hudson River power plants science and controversy: Barnthouse, L. W. "Impacts of Power Plant Cooling Systems on Estuarine Fish Populations: The Hudson River after 25 Years," *Environmental Science and Policy* 3 (2000): 341–48.

Chapter 10

117–18. international caviar market of late 1800s: Saffron, Inga. *The Strange History and Uncertain Future of the World's Most Coveted Delicacy* (New York: Broadway, 2002).

117–18. Delaware Bay sturgeon fishery: Carey, *The Philosopher Fish: Sturgeon, Caviar, and the Geography of Desire.*

118–20. importance of striped bass fisheries, their collapse, and restoration: Richards, R. A., and P. J. Rago. "A Case History of Effective Fishery Management: Chesapeake Bay Striped Bass," *North American Journal of Fisheries Management* 19 (1999): 356–75.

120. recovery requirements for striped bass at time of stock collapse: Boreman, J. "Production and Harvest of Anadromous Striped Bass Stocks Along the Atlantic Coast," *Transactions of the American Fisheries Society* 114 (1985): 3–7.

120–21. overview of "precautionary principle": Kriebel, D., and nine coauthors. "The Precautionary Principle in Environmental Science," *Environmental Health Perspectives* 109 (2001): 871–76.

121. "One observer wrote of elvers ascending streams": Gunther, Albert Carl Ludwig Gotthilf. *An Introduction to the Study of Fishes* (Edinburgh: Charles and Black, 1880).

121–22. newspaper account of eel elver migration: *Annual Report of the Commissioner of Fisheries.* New York State. Commissioners of Fisheries,

Game and Forests, Fourth Annual Report (New York and Albany: Wynkoop, Hallenbeck and Crawford, 1899).

122. estimates of eel declines in United States and Europe and conservation issues: Vélez-Espino, L. A., and M. A. Koops. "A Synthesis of the Ecological Processes Influencing Variation in Life History and Movement Patterns of American Eel: Towards a Global Assessment," *Reviews in Fish Biology and Fisheries* 20 (2010): 163–86; Feuneun, E. "Management and Restoration of European Eel Population (*Anguilla anguilla*): An Impossible Bargain," *Ecological Engineering* 18 (2002): 575–91.

122–23. eel elver fishery in Maine 2012: Goodnough, Abby. "Netting Tiny Eels and Big Profits," *New York Times,* March 29, 2012.

Chapter 11

124. David Brower and environmentalists' perspective on dams: McPhee, John. *Encounters with the Archdruid* (New York: Farrar, Straus and Giroux, 1971).

127. behavior of shad on Penobscot River after spawning migration blocked by dam: Maine Fisheries Commissioners Charles Atkins and Nathan Foster in 1869, as quoted in Watts, *Alewife: A Documentary History of the Alewife in Maine and Massachusetts.*

127–29. Overview of ecological effects of dams: Pringle, C. "What Is Hydrologic Connectivity and Why Is It Ecologically Important?" *Hydrological Processes* 17 (2003): 2685–89.

128. interrelationship among freshwater mussels, diadromous fish, and dams: Freeman, M. C., C. M. Pringle, E. A. Greathouse, and B. J. Freeman. "Ecosystem-level Consequences of Migratory Faunal Depletion Caused by Dams," *American Fisheries Society Symposium* 35 (2003): 255–66.

129. silica trapped behind dams: Humborg, C., V. Ittekkot, A. Cociasu, and B. v. Bodungen. "Effect of Danube River Dam on Black Sea Biogeochemistry and Ecosystem Structure," *Nature* 386 (1997): 385–88.

129. dams in the Hudson River watershed: Swaney, D. P., K. E. Limburg, and K. Stainbrook. "Some Historical Changes in the Patterns of

Population and Land Use in the Hudson Watershed," *American Fisheries Society Symposium* 51 (2006): 75–112.

129–30. loss of spawning habitat by shad: Limburg et al., *American Shad in its Native Range.*

130. loss of alewife spawning habitat and decreases in abundance in Maine: Hall, C. J., A. Jordaan, and M. G. Frisk. "The Historic Influence of Dams on Diadromous Fish Habitat with a Focus on River Herring and Hydrologic Longitudinal Connectivity," *Landscape Ecology* 26 (2011): 95–107.

130–31. eels in the St. Lawrence River: Casselman, *Dynamics of Resources of the American Eel.*

133–35. flaws in the flood-control paradigm: Black, Peter E. "The U.S. Flood Control Paradigm at 75: Environmental Issues," *Journal of the American Water Resources Association* 48 (2012): 244–55.

135–36. sediments behind Susquehanna River dams: Horton, Tom. "It's Time to Get Serious about Conowingo's Trapped Sediments," *Chesapeake Bay Journal* (May 2012).

137. the Johnstown flood: McCullough, David. *The Johnstown Flood* (New York: Simon & Schuster, 1968).

Chapter 12

141–42. Edith's checkerspot butterfly and climate change: Parmesan, Camille. "Climate and Species Range," *Nature* 382 (1996): 765–66.

142. climate change and species distribution changes: Parmesan, C., and G. Yohe. "A Globally Coherent Fingerprint of Climate Change Impacts Across Natural Systems," *Nature* 421 (2003): 37–42.

142. timing of spawning runs in Maine and climate change: Huntington, T. G., G. A. Hodgkins, and R. W. Dudley. "Historical Trend in River Ice Thickness and Coherence in Hydroclimatological Trends in Maine," *Climatic Change* 61 (2003): 217–36.

142–43. timing of salmon runs in Connecticut River: Juanes, F., S. Gephard, and K. F. Beland. "Long-Term Changes in Migration Timing of Adult Atlantic Salmon (*Salmo salar*) at the Southern End of the Species Distribution," *Canadian Journal of Fisheries and Aquatic Sciences* 61 (2004): 2392–2400.

143. timing of alewife runs in eastern Connecticut: Ellis, D., and J. C. Vokoun. "Earlier Spring Warming of Coastal Streams and Implications for Alewife Migration Timing," *North American Journal of Fisheries Management* 29 (2009): 1584–89.

143. outmigration of Allis shad in Loire River: Boisneau, C., F. Moatar, M. Bodin, and P. Boisneau. "Does Global Warming Impact on Migration Patterns and Recruitment of Allis Shad (*Alosa alosa* L.) Young of the Year in the Loire River, France?" *Hydrobiologia* 602 (2008): 179–86.

145–47. theory of low Atlantic salmon abundance in New England: Carlson, Catherine C. "Where's the Salmon? A Reevaluation of the Role of Anadromous Fisheries to Aboriginal New England," pages 47–80 in *Human Holocene Ecology in North America*. George P. Nicholas, ed. (New York: Plenum Press, 1988); Carlson, Catherine C. "The Significance of Atlantic Salmon," *Federal Archaeology* 8 (Fall/Winter 1996): 22–30.

148. additional archaeological research on salmon remains in Maine: Robinson, B. S., G. L. Jacobson, M. G. Yates, A. E. Spiess, and E. R. Cowie. "Atlantic Salmon, Archaeology and Climate Change in New England," *Journal of Archaeological Science* 36 (2009): 2184–91.

149. presence of salmon in Delaware River: Daniels, R. A., and D. Peteet. "Fish Scale Evidence for Rapid Post-Glacial Colonization of an Atlantic Coastal Pond," *Global Ecology & Biogeography Letters* 7 (1998): 467–76.

149–50. rainbow smelt and warming in Hudson River: Waldman, J. "The Diadromous Fish Fauna of the Hudson River: Life Histories, Conservation Concerns, and Research Avenues," pages 171–88 in J. S. Levinton and J. R. Waldman (eds.). *The Hudson River Estuary* (New York: Cambridge, 2006).

150. decline of rainbow smelt in Connecticut: Fried, H. A., and Schultz, E. T. "Anadromous Rainbow Smelt and Tomcod in Connecticut: Assessment of Populations, Conservation Status, and Need for Restoration

Plan" (2006). *EEB Articles.* Paper 18. http://digitalcommons.uconn.edu/eeb_articles/18.

153–54. Chesapeake Bay Anadromous and Shelf-Spawning Species theory: Wood, R. J., and H. M. Austin. "Synchronous Multidecadal Fish Recruitment Patterns in Chesapeake Bay, USA," *Canadian Journal of Fisheries and Aquatic Sciences* 66 (2009) 496–508.

155. recent and forecast sea level rise on US Eastern Seaboard: Sallenger Jr., A. H., K. S. Doran, and P. A. Howd. "Hotspot of Accelerated Sea-Level Rise on the Atlantic Coast of North America," *Nature Climate Change* 2 (2012): 884–88.

Chapter 13

158. fish movements to Hudson River via canals: Waldman, J. R., T. R. Lake, and R. E. Schmidt. "Biodiversity and Zoogeography of the Fishes of the Hudson River Watershed and Estuary," *American Fisheries Society Symposium* 51 (2006): 129–50.

159. "Storm of the Century" and sturgeon escapes into Gironde River: Maury-Bachet, R., E. Rochard, G. Durrieu, and A. Boudou. "The 'Storm of the Century' (December 1999) and the Accidental Escape of Siberian Sturgeon into the Gironde Estuary (Southwest France): An Original Approach for Metal Contamination," *Environmental Science and Pollution Research* 15 (2008) 89–94.

159–60. colonization of the Hudson River by zebra mussels: Strayer, D. L., J. Powell, P. Ambrose, L. C. Smith, M. L. Pace, and D. T. Fischer. "Arrival, Spread and Early Dynamics of a Zebra Mussel (*Dreissena polymorpha*) Population in the Hudson River Estuary," *Canadian Journal of Fisheries and Aquatic Sciences* 53 (1996): 1143–49.

160. changes to Hudson River fishes caused by zebra mussels: Strayer, D. L., K. A. Hattala, and A. W. Kahnle. "Effects of an Invasive Bivalve (*Dreissena polymorpha*) on Fish in the Hudson River Estuary," *Canadian Journal of Fisheries and Aquatic Sciences* 61 (2004): 924–41.

162. nonnative fishes in the Hudson River: Waldman, Lake, and Schmidt, "Biodiversity and Zoogeography of the Fishes of the Hudson River Watershed and Estuary."

162–63. nonnative fish in US East Coast rivers: Rinne, J. N., R. M. Hughes, and B. Calamusso, editors. *Historical Changes in Large River Fish Assemblages of the Americas.* American Fisheries Society Symposium 45 (Bethesda, MD: American Fisheries Society, 2005).

163–64. Asian carp threaten Great Lakes: Stokstad, E. "Biologists Rush to Protect Great Lakes from Onslaught of Carp," *Science* 327 (2010): 932.

164. effects of water chestnut in Hudson River: Hummel, M. and S. Findlay. "Effects of Water Chestnut (*Trapa natans*) Beds on Water Chemistry in the Tidal Freshwater Hudson River," *Hydrobiologia* 559 (2006): 169–81.

165. *Anguillicola crassus* parasite of eel: Barse, A. M., and D. H. Secor. "An Exotic Nematode Parasite of the American Eel," *Fisheries* 24 (2) (1999): 6–10.

Chapter 14

168. early observation of sturgeon in Merrimack River: Wood, *New England's Prospect.*

168. ocean landings of giant sturgeon: Collette and Klein-Macphee, *Bigelow and Schroeder's Fishes of the Gulf of Maine.*

168. archaeological findings on sturgeon from Jamestown, Virginia: Balazik, M. T., G. C. Garman, M. L. Fine, C. H. Hager, and S. P. McInich. "Changes in Age Composition and Growth Characteristics of Atlantic Sturgeon (*Acipenser oxyrinchus oxyrinchus*) Over 400 Years," *Biology Letters* 6 (2010): 708–10.

169. large shad from the Susquehanna River: Gerstell, *American Shad in the Susquehanna River Basin.*

169–70. records of giant Atlantic salmon: Buller, F. *The Domesday Book of Giant Salmon* (London: Constable, 2007).

170–71. account of capture of record striped bass by Charles Church: Church, Charles B. "Taking the Record Striped Bass of 1913," as told by Charles B. Church in his own words.

171–72. historical records of large striped bass: Karas, Nick. *The Striped Bass* (New York: Lyons Press, 1993).

173–74. laboratory experiments on harvest and growth with silversides and "Darwinian debt": Walsh, M. R., S. B. Munch, S. Chiba, and D. O. Conover. "Maladaptive Changes in Multiple Traits Caused by Fishing: Impediments to Population Recovery," *Ecology Letters* 9 (2006): 142–48.

174–75. effects of warming seas on fish growth: Cheung, W. W. L., and seven coauthors. "Shrinking of Fishes Exacerbates Impacts of Global Ocean Changes on Marine Ecosystems," *Nature Climate Change* 3 (2013): 254–58.

Chapter 15

176. solitary salmon in Connecticut River fisherman's net: Netboy, Anthony. *The Salmon: Their Fight for Survival* (Boston: Houghton Mifflin, 1973).

178. plummeting worldwide catches of Atlantic salmon: Parfit, Michael. "Lost at Sea," *Smithsonian* (April 2002): 68–77.

179. marine movements of Atlantic salmon: Dadswell, M. J., A. D. Spares, J. M. Reader, and M. J. W. Stokesbury. "The North Atlantic Subpolar Gyre and the Marine Migration of Atlantic Salmon, *Salmo salar:* The 'Merry-Go-Round' Hypothesis," *Journal of Fish Biology* 77 (2010): 435–67.

180–81. Orri Vigfússon: North Atlantic Salmon Fund, www .nasfworldwide.com.

181. escapes of farmed Atlantic salmon in Maine: Kircheis, F. *Annual Report to the Maine Legislature Fish and Wildlife Committee for the Period January through December 2001.* Maine Atlantic Salmon Commission, Augusta, Maine, 2001.

182. general effects of aquacultured salmon escapees on wild salmon, including disease transmission: Naylor, R. and nine coauthors. "Fugitive

Salmon: Assessing the Risks of Escaped Fish from Net-Pen Aquaculture," *Bioscience* 55 (2005): 427–37.

182–83. link between river herring and cod stocks: Jordaan, A., C. Hall, and M. Frisk. "Is the Recovery of Cod (*Gadus morhua*) along the Maine Coast Limited by Reduced Anadromous River Herring Populations?" *Final Report to Mia J. Tegner Memorial Research Grant in Marine Historical Ecology and Environmental History.* October 2008.

183–84. old-time eel fishing in Moriches Bay, Long Island, New York: "Eels and Eel Catchers," page 466 in *The New England Farmer* (Boston: Joel Nourse, 1856).

184. 1938 fish survey of waters of Long Island, New York: Greeley, J. R. "A Biological Survey of the Fresh Waters of Long Island," *28th Annual Report New York State Conservation Department,* 1938.

184–85. eel recruitment and climate: Knights, B. "A Review of the Possible Impacts of Long-Term Oceanic and Climate Changes and Fishing Mortality on Recruitment of Anguillid Eels of the Northern Hemisphere," *The Science of the Total Environment* 310 (2003): 237–44.

Chapter 16

187–88. history of Tocks Island Dam project: Albert, R. C. *Damming the Delaware: The Rise and Fall of Tocks Island Dam* (University Park: Penn State University Press, 1990).

191–92. Robert Juet and "salmon": Waldman, John. *Heartbeats in the Muck: The History, Sea Life, and Environment of New York Harbor* (New York: Lyons Press, 1999).

195–97. haul seining for striped bass on Long Island, New York: Cole, John N. *Striper: A Story of Fish and Man* (New York: Little, Brown, 1978); Matthiessen, *Men's Lives.*

197–98. comparison of wild and hatchery-reared striped bass: Waldman, J. R., and V. J. Vecchio. "Selected Biocharacteristics of Hatchery-Reared Striped Bass Captured in New York Ocean Waters," *North American Journal of Fisheries Management* 16 (1996): 14–23.

199–200. striped bass history and restoration in Gulf of Mexico rivers: Long, E. A., C. L. Mesing, K. J. Harrington, R. R. Weller, and I. I. Wirgin. "Restoration of Gulf Striped Bass: Lessons and Management Implications," *American Fisheries Society Symposium* 80 (2013): 25–64.

Chapter 17

204–5. environmental history of Naugatuck River: Black, George. *The Trout Pool Paradox* (Boston: Houghton Mifflin Harcourt, 2004). See also Naugatuck River Restoration: www.naugawatshed.org.

205–6. Interior Secretary Bruce Babbitt on dams and dam removals: Babbitt, Bruce. "What Goes Up, May Come Down," *Bioscience* 52 (2002): 656–58.

209–10. quote on impossibility of restoring alewife to Mill River: Belding, David, L. "A Report Upon the Alewife Fisheries of Massachusetts," *Division of Fisheries and Game, Department of Conservation,* Boston, 1921.

Chapter 18

218. Turners Falls Massacre: Schultz, E. B., and M. J. Tougias. *King Philip's War: The History and Legacy of America's Forgotten Conflict* (Woodstock, VT: Countryman Press, 2000).

219–20. quote by John Hay on fish ladders: Hay, *The Run.*

224–28. program to restore salmon and other diadromous fishes on Connecticut River: Gephard, S., and J. R. McMenemy. "An Overview of the Program to Restore Atlantic Salmon and Other Diadromous Fishes to the Connecticut River with Notes on the Current Status of These Species in the River," *American Fisheries Society Monograph* 9 (2004): 287–317.

Chapter 19

238–39. recovery of striped bass fishery: Richards and Rago, *A Case History of Effective Fishery Management: Chesapeake Bay Striped Bass.*

244. increase in Hudson River shortnose sturgeon population: Bain, M. and five coauthors. "Recovery of a US Endangered Fish," *PLoS ONE* 2(1): e168. doi:10.1371/journal.pone.0000168.

245. Penobscot River restoration program: Carpenter, Murray. "Dam Removal to Help Restore Spawning Grounds," *New York Times,* June 12, 2012.

247–51. recent conservation efforts and water quality issues on Blackstone River: Williams, Ted. "Clean Water on Hold," *Fly Rod and Reel* (January 2013): 14–20.

Chapter 20

254–55. selection effects of a fish ladder: Haugen, T. O., P. Aass, N. C. Stenseth, and L. A. Vøllestad. "Changes in Selection and Evolutionary Responses in Migratory Brown Trout Following the Construction of a Fish Ladder," *Evolutionary Applications* 1 (2008): 319–35.